AQA Science

Exclusively endorsed and approved by AQA

Patrick Fullick

Series Editor: Lawrie Ryan

GCSE Chemistry

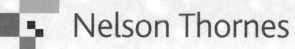
Nelson Thornes

JULIA COELITO

Published in 2006 by:
Nelson Thornes Ltd
Delta Place
27 Bath Road
CHELTENHAM
GL53 7TH
United Kingdom

10 11 12 13 / 15 14 13 12 11 10 9

A catalogue record for this book is available from the British Library

ISBN 978 0 7487 9644 1

Cover photographs: flames by Photodisc 29 (NT); chemical crystals by Photodisc 4 (NT); chemical flasks by Photodisc 54 (NT)
Cover bubble illustration by Andy Parker
Illustrations by Bede Illustration, Beverly Curl, Kevin Jones Associates and Roger Penwill
Page make-up by Wearset Ltd

Printed in China by 1010 Printing international Ltd

GCSE Chemistry · Contents

How to use this book

This textbook will help you throughout your GCSE course and to prepare for AQA's exams. It is packed full of features to help you to achieve the best result you can.

> Some of the text is in a box marked HIGHER. You have to include these parts of the book if you are taking the Higher Tier exam. If you are taking the Foundation Tier exam, you can miss these parts out.
>
> The same applies to any Learning Objectives, Key Points or Questions marked [Higher].

HIGHER

a) What are the yellow boxes?

To check you understand the science you are learning, questions are integrated into the main text. The information needed to answer these is on the same page, so you don't waste your time flicking through the entire book.

LEARNING OBJECTIVES

By the end of the lesson you should be able to answer the questions posed in the learning objectives; if you can't, review the content until it's clear.

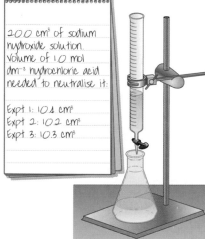

Figure 1 Key diagrams are as important as the text. Make sure you use them in your learning and revision.

Key words
Important scientific terms are shown like this:

observation or **anomalous**

You can look up the words shown like this – **bias** – in the glossary.

GET IT RIGHT!
Avoid common mistakes and gain marks by sticking to this advice.

KEY POINTS

If you remember nothing else, remember these! Learning the key points in each lesson is a good start. They can be used in your revision and help you summarise your knowledge.

PRACTICAL

Become familiar with key practicals. A simple diagram and questions make this feature a short introduction, reminder or basis for practicals in the classroom.

E.g.
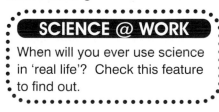
1 mol/dm³

Weak acid

DID YOU KNOW?
Curious examples of scientific points that are out of the ordinary, but true…

FOUL FACTS
Some science is just too gruesome to ignore. Delve into the horrible yet relevant world of Foul Facts.

At the start of each unit you will find a double-page introduction. This reminds you of ideas from previous work that you will need. The recap questions and activity will help find out if you need some revision before starting.

SCIENCE @ WORK
When will you ever use science in 'real life'? Check this feature to find out.

SUMMARY QUESTIONS

Did you understand everything? Get these questions right, and you can be sure you did. Get them wrong, and you might want to take another look.

The ideas in 'How Science Works' are covered in the first chapter. You will need to refer back to this chapter as you work through the course.

This first chapter looks at 'How Science Works'. It is an important part of your GCSE because the ideas introduced here will crop up throughout your course. You will be expected to collect scientific evidence and to understand how we use evidence. These concepts will be assessed as the major part of your internal school assessment. You will take one or more 45-minute tests on data you have collected previously plus data supplied for you in the test. These are called Investigative Skills Assignments. The ideas in 'How Science Works' will also be assessed in your examinations.

What you already know

Here is a quick reminder of previous work with investigations that you will find useful in this chapter:

- You will have done some practical work and know how important it is to keep yourself and others safe.
- Before you start investigating you usually make a prediction, which you can test.
- Your prediction and plan will tell you what you are going to change and what you are going to measure.
- You will have thought about controls.
- You will have thought about repeating your readings.
- During your practical work you will have written down your results, often in a table.
- You will have plotted graphs of your results.
- You will have made conclusions to explain your results.
- You will have thought about how you could improve your results, if you did the work again.

RECAP QUESTIONS

Ed wrote this account of a practical he did:

I wanted to find out how much sugar would dissolve in water at different temperatures. I thought that the hotter the water the more sugar would dissolve.

So I took 100cm³ of tap water, measured its temperature (18°C) and stirred in as much sugar as I could. I dissolved 32 grams. I then took the same amount of water out of the fridge, the water was at 8°C. This time only 10 grams dissolved. I did exactly the same with 100cm³ of water at 25°C and at 45°C. The first one of these dissolved 50 grams. The second one dissolved 95 grams.

1 What was Ed's prediction?

2 What was the variable he chose to change? (We call this the independent variable.)

3 What was the variable he measured to judge the effect of varying the independent variable? (We call this the dependent variable. Its value **depends** on the value chosen for the independent variable.)

4 Write down a variable that Ed controlled.

5 Write down a variable that Ed did not say he had controlled.

6 Make a table of Ed's results.

7 Draw a graph of his results.

8 Write a conclusion for Ed.

9 How do you think Ed could have improved his results?

How science works for us

Science works for us all day, every day. You do not need to know how a mobile phone works to enjoy sending text messages. But, think about how you started to use your mobile phone or your television remote control. Did you work through pages of instructions? Probably not!

You knew that pressing the buttons would change something on the screen (**knowledge**). You played around with the buttons, to see what would happen (**observation**). You had a guess at what you thought might be happening (**prediction**) and then tested your idea (**experiment**).

If your prediction was correct you remembered that as a **fact**. If you could repeat the operation and get the same result again then you were very pleased with yourself. You had shown that your results were **reliable**.

Working as a scientist you will have knowledge of the world around you and particularly about the subject you are working with. You will observe the world around you. An enquiring mind will then lead you to start asking questions about what you have observed.

Science moves forward by slow steady steps. When a genius such as Einstein comes along then it takes a giant leap. Those small steps build on knowledge and experience that we already have.

Each small step is important in its own way. It builds on the body of knowledge that we have. In 1675 a German chemist tried to extract gold from urine. He must have thought that there was a connection between the two colours. He was wrong, but after a long while, with an incredible stench coming from his laboratory, the urine began to glow.

He had discovered phosphorus. A Swedish scientist worked out how to manufacture phosphorus without the smell of urine. Phosphorus catches fire easily. That is why most matches these days are manufactured in Sweden.

Thinking scientifically

Figure 1 Discussing fireworks

ACTIVITY

Now look at Figure 1 with your scientific brain.

- Fireworks must be safe to light. Therefore you need a fuse that will last long enough to give you time to get well out of the way.
- Fuses can be made by dipping a special type of cotton into a mixture of two chemicals. One chemical (A) reacts by burning, the other (B) doesn't.
- The chemicals stick to the cotton. Once it is lit, the cotton will continue to burn, setting the firework off. The concentrations of the two chemicals will affect how quickly the fuse burns.

In groups discuss how you could work out the correct concentrations of the chemicals to use. You want the fuse to last long enough to get out of the way, but not to burn so long that we all get bored waiting for the firework to go off!

You can use the following headings to discuss your investigation. One person should be writing your ideas down, so that you can discuss them with the rest of the class.

- What prediction can you make about the concentration of the two chemicals (A and B) and the fuse?
- What would be your independent variable?
- What would be your dependent variable?
- What would you have to control?
- Write a plan for your investigation.
- How could you make sure your results were reliable?

H2

Fundamental ideas about how science works

LEARNING OBJECTIVES

1 How do you spot when a person has an opinion that is not based on good science?
2 What is the importance of continuous, ordered and categoric variables?
3 What is meant by reliable evidence and valid evidence?
4 How can two sets of data be linked?

NEXT TIME YOU...

... read a newspaper article or watch the news on TV ask yourself if that research is valid and reliable. (See page 5.) Ask yourself if you can trust the opinion of that person.

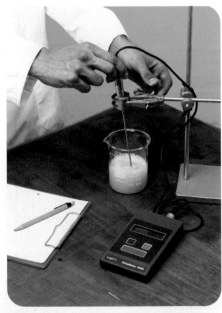

Figure 1 Student recording a range of temperatures – an example of a continuous variable

Science is too important for us to get it wrong

Sometimes it is easy to spot when people try to use science poorly. Sometimes it can be funny. You might have seen adverts claiming to give your hair 'body' or sprays that give your feet 'lift'!

On the other hand, poor scientific practice can cost lives.

Some years ago a company sold the drug thalidomide to people as a sleeping pill. Research was carried out on animals to see if it was safe. The research did not include work on pregnant animals. The opinion of the people in charge was that the animal research showed the drug could be used safely with humans.

Then the drug was also found to help ease morning sickness in pregnant women. Unfortunately, doctors prescribed it to many women, resulting in thousands of babies being born with deformed limbs. It was far from safe.

These are very difficult decisions to make. You need to be absolutely certain of what the science is telling you.

a) Why was the opinion of the people in charge of developing thalidomide based on poor science?

Deciding on what to measure

You know that you have an independent and a dependent variable in an investigation. These variables can be one of four different types:

● A **categoric variable** is one that is best described by a label (usually a word). The type of gas given off in a reaction is a categoric variable, e.g. the gas given off was carbon dioxide gas.
● A **discrete variable** is one that you describe in whole numbers. The number of lumps of marble chip used.
● An **ordered variable** is one where you can put the data into order, but not give it an actual number. The amount of gas given off in one reaction compared to another is an ordered variable, e.g. more carbon dioxide was given off by the hydrochloric acid.
● A **continuous variable** is one that we measure, so its value could be any number. Volume of gas given off (as measured by a gas syringe or an upturned measuring cylinder of water) is a continuous variable, e.g. $45\,cm^3$; $43\,cm^3$; $37\,cm^3$; $56\,cm^3$; $33\,cm^3$ of carbon dioxide were given off.

When designing your investigation you should always try to measure continuous data whenever you can. This is not always possible, so you should then try to use ordered data. If there is no other way to measure your variable then you have to use a label (categoric variable).

b) Imagine you were testing the heat given out in three different reactions (A, B and C). Would it be best to say i) reactions A and B felt warm, but C felt hot, or ii) reaction C got hottest, followed by A and and finally B, or iii) the rise in temperature in reaction C was 31°C, in A it was 16°C and in B it was 14°C?

Making your investigation reliable and valid

When you are designing an investigation you must make sure that others can get the same results as you – this makes it **reliable**.

You must also make sure you are measuring the actual thing you want to measure. If you don't, your data can't be used to answer your original question. This seems very obvious but it is not always quite so easy. You need to make sure that you have **controlled** as many other variables as you can, so that no-one can say that your investigation is not **valid**. A valid investigation should be reliable *and* answer the original question.

c) State one way in which you can show that your results are valid.

Figure 2 The aftermath of an earthquake

How might an independent variable be linked to a dependent variable?

Variables can be linked together for one of three reasons:

● It could be because one variable has caused a change in the other, e.g. the higher the temperature the quicker the glue will set. This is a **causal link**.
● It could be because a third variable has caused changes in the two variables you have investigated, e.g. the denser iron ore is, the more valuable it is. This is because there is an **association** between the two variables. Both of the variables are caused by the increased proportion of iron in the rock.
● It could be due simply to **chance**, e.g. a link between the number of deaths and the strength of an earthquake. (An earthquake in a built-up area may be weak but still cause many deaths – the link was just by chance.)

d) Describe a causal link that you have seen in chemistry.

SUMMARY QUESTIONS

1 Students were asked to find the solubility of three different solids – D, E and F.

Name each of the following types of dependent variable described by the students in a), b) and c):

a) D and E were 'soluble', whereas F was 'very soluble'.
b) F was most soluble, D was second and E was least soluble.
c) 59.8 g of F dissolved in 100 cm³ of water, 30.2 g of D dissolved in 100 cm³ of water, 25.9 g of F dissolved in 100 cm³ of water.

2 Some people believe that the artificial sweetener aspartame causes headaches and dizziness. Do you trust these opinions? What would convince you not to use aspartame?

KEY POINTS

1 Be on the lookout for non-scientific opinions.
2 Continuous data give more information than other types of data.
3 Check that evidence is reliable and valid.
4 Be aware that just because two variables are related it does not mean that there is a causal link between them.

H3 Starting an investigation

Observation

As humans we are sensitive to the world around us. We can use our many senses to detect what is happening. As scientists we use observations to ask questions. We can only ask useful questions if we know something about the observed event. We will not have all of the answers, but we know enough to start asking the correct questions.

If we observe that the weather has been hot today, we would not ask if it was due to global warming. If the weather was hotter than normal for several years then we could ask that question. We know that global warming takes many years to show its effect.

When you are designing an investigation you have to observe carefully which variables are likely to have an effect.

a) Would it be reasonable to ask if the iron in Figure 1 is rusting because of acid rain? Discuss your answer.

Amjid noticed that the driveway up to his house had cracks in the concrete on the left side of the driveway (observation). He was concerned because the driveway had only been laid for ten weeks. The work had been done last December. Before the builder came to look at it, Amjid thought of a few ideas to put to the builder.

- Did he have the correct amount of water in the concrete?
- Did he use the correct amount of cement?
- Could it be the car that was causing the damage?
- Did the builder dig the foundations deep enough?
- Did the builder put the same depth of foundations on both sides?
- Could the frost have caused the damage?
- Could the bushes growing next to the drive have caused the problem?

b) Discuss all of these good ideas and choose three that are the most likely.

Observations, backed up by really creative thinking and good scientific knowledge can lead to a **hypothesis**.

What is a hypothesis?

A hypothesis is a 'great idea'. Why is it so great? – well because it is a great observation that has some really good science to try to explain it.

For example, you observe that small, thinly sliced chips cook faster than large, fat chips. Your hypothesis could be that the small chips cook faster because the heat from the oil has a shorter distance to travel before it gets to the potato in the centre of the chips.

c) Check out the photograph of the rusting object in Figure 1 and spot anything that you find interesting. Use your knowledge and some creative thought to suggest a hypothesis for each observation you can make.

When making hypotheses you can be very imaginative with your ideas. However, you should have some scientific reasoning behind those ideas so that they are not totally bizarre.

Remember, your explanation might not be correct, but you think it is. The only way you can check out your hypothesis is to make it into a prediction and then test it by carrying out an investigation.

Observation ✛ knowledge ➡ hypothesis ➡ prediction ➡ investigation

Figure 1 Rusting lock

Starting to design a valid investigation

An investigation starts with a prediction. You, as the scientist, predict that there is a relationship between two variables.

- An **independent variable** is one that is changed or selected by you, the investigator.

- A **dependent variable** is measured for each change in your independent variable.

- All other variables become *control variables*, kept constant so that your investigation is a fair test.

If your measurements are going to be accepted by other people then they must be valid. Part of this is making sure that you are really measuring the effect of changing your chosen variable. For example, if other variables aren't controlled properly, they might be affecting the data collected.

Figure 2 Darren investigating the temperature change

d) Look at Figure 2. Darren was investigating the temperature change when adding anhydrous copper sulfate to water. He used a test tube for the reaction. What is wrong here?

SUMMARY QUESTIONS

1 Copy and complete using the words below:

    controlled   dependent   hypothesis   independent
                knowledge   prediction

Observations when supported by scientific …… can be used to make a …… . This can be the basis for a …… . A prediction links an …… variable to a …… variable. Other variables need to be …… .

2 Explain the difference between a hypothesis and a prediction.

KEY POINTS

1 Observation is often the starting point for an investigation.
2 Hypotheses can lead to predictions and investigations.
3 You must design investigations that produce valid results if you are to be believed.

H4 Building an investigation

Figure 1 Different types of paint

Fair testing

A **fair test** is one in which only the independent variable affects the dependent variable. All other variables are controlled.

This is easy to set up in the laboratory, but almost impossible in fieldwork. Imagine you were studying the quality of the water from wells at different distances from a factory. To make your investigation valid, you would want to take many samples from similar rock formations. You would choose sites where all of the many variables change in much the same way, except for the one you are investigating.

a) How would you set up an investigation to see how exposure to different amounts of sunlight affected different types of paint?

If you are investigating two variables in a large population then you will need to do a survey. Again it is impossible to control all of the other variables. Imagine you were investigating the effect of salt on blood pressure. You would have to choose people of the same age and same family history to test. The larger the sample size you test, the more reliable your results will be.

Control groups are used in investigations to try to make sure that you are measuring the variable that you intend to measure. When investigating the effects of a new drug, the control group will be given a placebo.

The control group think they are taking a drug but the placebo does not contain the drug. This way you can control the variable of '*thinking* that the drug is working' and separate out the effect of the actual drug.

Choosing values of a variable

Trial runs will tell you a lot about how your early thoughts are going to work out.

Do you have the correct conditions?
Suppose you are finding out how much gas was produced when different masses of a chemical were added to some water. Only very small amounts of gas were produced.

- There might not have been enough chemical added.
- It might not have been left long enough.

Have you chosen a sensible range?
If there is enough gas produced, but the results all look about the same:

- you might not have chosen a wide enough range of masses of the chemical.

Have you got enough readings that are close together?
If the results are very different from each other:

- you might not see a pattern if you have large gaps between readings over the important part of the range.

Accuracy

Accurate measurements are very close to the *true value*.

Your investigation should provide data that is accurate enough to answer your original question.

However, it is not always possible to know what that true value is.

How do you get accurate data?
- You can repeat your measurements and your mean is more likely to be accurate.
- Try repeating your measurements with a different instrument and see if you get the same readings.
- Use high quality instruments that measure accurately.
- The more carefully you use the measuring instruments, the more accuracy you will get.

Precision and reliability

If your repeated measurements are closely grouped together then you have precision and you have improved the reliability of your data.

Your investigation must provide data with sufficient precision. It's no use measuring the time for a fast reaction to finish using the seconds hand on a clock! If there are big differences within sets of repeat readings, you will not be able to make a valid conclusion. You won't be able to trust your data!

How do you get precise and reliable data?
- You have to use measuring instruments with sufficiently small scale divisions.
- You have to repeat your tests as often as necessary.
- You have to repeat your tests in exactly the same way each time.

A word of caution!

Be careful though – just because your results show precision does not mean your results are accurate. Look at the box opposite.

b) Draw a thermometer scale showing 4 results that are both accurate and precise.

The difference between accurate and precise results

Imagine measuring the temperature after a set time when a fuel is used to heat a fixed volume of water. Two students repeated this experiment, four times each. Their results are marked on the thermometer scales below:

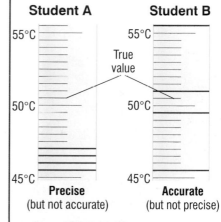

- A precise set of repeat readings will be grouped closely together.
- An accurate set of repeat readings will have a mean (average) close to the true value.

SUMMARY QUESTIONS

1 Copy and complete using the following terms:

      range   repeat   conditions   readings

    Trial runs give you a good idea of whether you have the correct ......; whether you have chosen the correct ......; whether you have enough ......; if you need to do ...... readings.

2 Use an example to explain how a set of repeat measurements could be accurate, but not precise.

3 Briefly describe how you would go about setting up a fair test in a laboratory investigation. Give your answer as general advice.

KEY POINTS

1 Care must be taken to ensure fair testing – as far as is possible.
2 You can use a trial run to make sure that you choose the best values for your variables.
3 Careful use of the correct equipment can improve accuracy.
4 If you repeat your results carefully they are likely to become more reliable.

H5 | Making measurements

LEARNING OBJECTIVES

1 Why do results always vary?
2 How do you choose instruments that will give you accurate results?
3 What do we mean by the sensitivity of an instrument?
4 How does human error affect results and what do you do with anomalies?

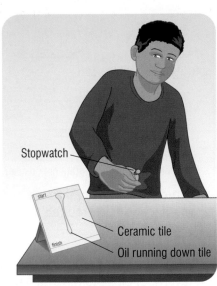

Figure 1 Student reading the arrival of the oil

DID YOU KNOW?

Professor Hough was investigating possible uses of sucrose (ordinary sugar) in industry. He had created a molecule of sucrose with three atoms of chlorine in it. He asked his new assistant to 'test' it. His assistant thought he had said 'taste' it. Fortunately for his assistant it did him no harm, but he noticed how incredibly sweet it was – a thousand times sweeter than sugar!

Using instruments

Do not panic! You cannot expect perfect results.

Try measuring the temperature of a beaker of water using a digital thermometer. Do you always get the same result? Probably not. So can we say that any measurement is absolutely correct?

In any experiment there will be doubts about actual measurements.

a) Look at Figure 1. Suppose, like this student, you tested the time it takes for one type of oil to flow down the tile. It is unlikely that you would get two readings exactly the same. Discuss all the possible reasons why.

When you choose an instrument you need to know that it will give you the accuracy that you want. That is, it will give you a true reading.

If you have used an electric water bath, would you trust the temperature on the dial? How do you know it is the true temperature? You could use a very expensive thermometer to calibrate your water bath. The expensive thermometer is more likely to show the true temperature. But can you really be sure it is accurate?

You also need to be able to use an instrument properly.

b) In Figure 1 the student is measuring the time it takes for the oil to reach the line. Why is the student unlikely to get a true measurement?

When you choose an instrument you need to decide how accurate you need it to be. Instruments that measure the same thing can have different sensitivities. The **sensitivity** of an instrument refers to the smallest change in a value that can be detected. This determines the precision of your measurements.

Choosing the wrong scale can cause you to miss important data or make silly conclusions, for example 'The amount of gold was the same in the two rings – they both weighed 5 grams.'

c) Match the following weighing machines to their best use:

Used to measure	Sensitivity of weighing machine
Cornflakes delivered to a supermarket	milligrams
Carbohydrate in a packet of cornflakes	grams
Vitamin D in a packet of cornflakes	micrograms
Sodium chloride in a packet of cornflakes	kilograms

Errors

Even when an instrument is used correctly, the results can still show differences.

Results may differ because of **random error**. This is most likely to be due to a poor measurement being made. It could be due to not carrying out the method consistently.

The error might be a **systematic error**. This means that the method was carried out consistently but an error was being repeated.

Check out these two sets of data that were taken from the investigation that Mark did. He tested 5 different oils. The third line is the time calculated from knowing the viscosity of the different oils:

Type of oil used	a	b	c	d	e
Time taken to flow down tile (seconds)	23.2	45.9	49.5	62.7	75.9
	24.1	36.4	48.7	61.5	76.1
Calculated time (seconds)	18.2	30.4	42.5	55.6	70.7

d) Discuss whether there is any evidence for random error in these results.
e) Discuss whether there is any evidence for systematic error in these results.

Anomalies

Anomalous results are clearly out of line. They are not those that are due to the natural variation you get from any measurement. These should be looked at carefully. There might be a very interesting reason why they are so different. If they are simply due to a random error, then they should be discarded (rejected).

If anomalies can be identified while you are doing an investigation, then it is best to repeat that part of the investigation.

If you find anomalies after you have finished collecting data for an investigation, then they must be discarded.

DID YOU KNOW?

Ludwig Mond was a very important industrial chemist. He was passing carbon monoxide over heated nickel powder. One evening after experimenting with this apparatus, his assistant noticed that as the equipment cooled it left a green flame rather than the usual blue flame. Mond really couldn't understand this and rather than ignoring this anomaly started to investigate. He eventually discovered that it was nickel carbonyl. This gave rise to a new way of extracting nickel from nickel ore and a whole new industry which was set up in Swansea.

SUMMARY QUESTIONS

1 Copy and complete using the words below:

   accurate   discarded   random   sensitivity   systematic   use   variation

   There will always be some …… in results. You should always choose the best instruments that you can to get the most …… results. You must know how to …… the instrument properly. The …… of an instrument refers to the smallest change that can be detected. There are two types of error – …… and ……. Anomalies due to random error should be ……

2 Which of the following will lead to a systematic error and which to a random error?
   a) Using a weighing machine, which has something stuck to the pan on the top.
   b) Forgetting to re-zero the weighing machine.

KEY POINTS

1 Results will nearly always vary.
2 Better instruments give more accurate results.
3 Sensitivity of an instrument refers to the smallest change that it can detect.
4 Human error can produce random and systematic errors.
5 We examine anomalies; they might give us some interesting ideas. If they are due to a random error, we repeat the measurements. If there is no time to repeat them, we discard them.

H6 | Presenting data

Figure 1 Student using an oxygen meter

For this section you will be working with data from this investigation:

Mel took a litre (1 dm³) of tap water. She shook it vigorously for exactly 2 minutes. She tried to get as much oxygen to dissolve in it as possible.

Then she took the temperature of the water. She immediately tested the oxygen concentration, using an oxygen meter.

Tables

Tables are really good for getting your results down quickly and clearly. You should design your table **before** you start your investigation.

Your table should be constructed to fit in all the data to be collected. It should be fully labelled, including units.

In some investigations, particularly fieldwork, it is useful to have an extra column for any notes you might want to make as you work.

While filling in your table of results you should be constantly looking for anomalies.

● Check to see if a repeat is sufficiently close to the first reading.
● Check to see if the pattern you are getting as you change the independent variable is what you expected.

Remember a result that looks anomalous should be checked out to see if it really is a poor reading or if it might suggest a different hypothesis.

Planning your table

Mel knew the values for her independent variable. We always put these in the first column of a table. The dependent variable goes in the second column. Mel will find its values as she carries out the investigation.

So she could plan a table like this:

Temperature of water (°C)	Concentration of oxygen (mg/dm³)
5	
10	
16	
20	
28	

Or like this:

Temperature of water (°C)	5	10	16	20	28
Concentration of oxygen (mg/dm³)					

All she had to do in the investigation was to write the correct numbers in the second column to complete the top table.

Mel's results are shown in the alternative format in the table below:

Temperature of water (°C)	5	10	16	20	28
Concentration of oxygen (mg/dm³)	12.8	11.3	9.9	9.1	7.3

The range of the data

Pick out the maximum and the minimum values and you have the range. You should always quote these two numbers when asked for a range. For example, the range is between .... (the lowest value) and .... (the highest value) – and don't forget to include the units!

a) What is the range for the independent variable and for the dependent variable in Mel's set of data?

The mean of the data

Often you have to find the mean of each repeated set of measurements.

You add up the measurements in the set and divide by how many there are. Miss out any anomalies you find.

The repeat values and mean can be recorded as shown below:

Temperature of water (°C)	Concentration of oxygen (mg/dm³)			
	1st test	2nd test	3rd test	Mean

Displaying your results

Bar charts

If you have a categoric or an ordered independent variable and a continuous dependent variable then you should use a bar chart.

Line graphs

If you have a continuous independent and a continuous dependent variable then a line graph should be used.

Scatter graphs or scattergrams

Scatter graphs are used in much the same way as line graphs, but you might not expect to be able to draw such a clear line of best fit. For example, if you wanted to see if people's lung capacity was related to how long they could hold their breath, you would draw a scatter graph with your results.

SUMMARY QUESTIONS

1 Copy and complete using the words below:

categoric   continuous   mean   range

The maximum and minimum values show the ...... of the data. The sum of all the values divided by the total number of the values gives the .... Bar charts are used when you have a ...... independent variable and a continuous dependent variable.
Line graphs are used when you have ...... independent and dependent variables.

2 Draw a graph of Mel's results from the top of this page.

NEXT TIME YOU...

... make a table for your results remember to include:
- headings,
- units,
- a title.

... draw a line graph remember to include:
- the independent variable on the x-axis,
- the dependent variable on the y-axis,
- a line of best fit,
- labels, units and a title.

GET IT RIGHT!

Marks are often dropped in the ISA by candidates plotting points incorrectly. Also use a **line of best fit** where appropriate – don't just join the points 'dot-to-dot'!

KEY POINTS

1 The range states the maximum and the minimum values.
2 The mean is the sum of the values divided by how many values there are.
3 Tables are best used during an investigation to record results.
4 Bar charts are used when you have a categoric or an ordered independent variable and a continuous dependent variable.
5 Line graphs are used to display data that are continuous.

H7 | Using data to draw conclusions

Identifying patterns and relationships

Now that you have a bar chart or a graph of your results you can begin to look for patterns. You must have an open mind at this point.

Firstly, there could still be some anomalous results. You might not have picked these out earlier. How do you spot an anomaly? It must be a significant distance away from the pattern, not just within normal variation.

A line of best fit will help to identify any anomalies at this stage. Ask yourself – do the anomalies represent something important or were they just a mistake?

Secondly, remember a line of best fit can be a straight line or it can be a curve – you have to decide from your results.

The line of best fit will also lead you into thinking what the relationship is between your two variables. You need to consider whether your graph shows a **linear** relationship. This simply means, can you be confident about drawing a straight line of best fit on your graph? If the answer is yes – then is this line positive or negative?

a) Say whether graphs (i) and (ii) in Figure 1 show a positive or a negative linear relationship.

Look at the graph in Figure 2. It shows a positive linear relationship. It also goes through the origin (0,0). We call this a **directly proportional** relationship.

Your results might also show a curved line of best fit. These can be predictable, complex or very complex! Look at Figure 3 below.

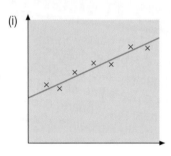

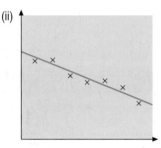

Figure 1 Graphs showing linear relationships

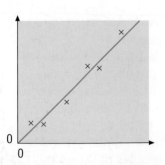

Figure 2 Graph showing a directly proportional relationship

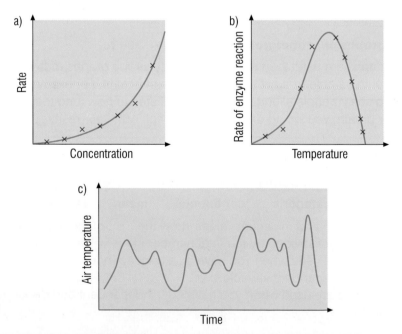

Figure 3 a) Graph showing predictable results. b) Graph showing complex results. c) Graph showing very complex results.

Drawing conclusions

Your graphs are designed to show the relationship between your two chosen variables. You need to consider what that relationship means for your conclusion.

There are three possible links between variables. (See page 5.) They can be:

- causal,

- due to association, or

- due to chance.

You must decide which is the most likely. Remember a positive relationship does not always mean a causal link between the two variables.

Poor science can often happen if a wrong decision is made here. Newspapers have said that living near electricity sub-stations can cause cancer. All that scientists would say is that there is possibly an association. Getting the correct conclusion is very important.

You will have made a prediction. This could be supported by your results. It might not be supported or it could be partly supported. Your results might suggest some other hypothesis to you.

Your conclusion must go no further than the evidence that you have. For example, your results might show that if you double the concentration of a reactant, you double the rate of reaction. However, we can't be certain this relationship holds true beyond the range of concentrations investigated. Further tests would be required.

Evaluation

If you are still uncertain about a conclusion, it might be down to the reliability and the validity of the results. You could check these by:

- looking for other similar work on the Internet or from others in your class,

- getting somebody else to re-do your investigation, or

- trying an alternative method to see if you get the same results.

SUMMARY QUESTIONS

1 Copy and complete using the words below:

   anomalous   complex   directly   negative   positive

   Lines of best fit can be used to identify …… results. Linear relationships can be …… or …… . If a graph goes through the origin then the relationship could be …… proportional. Often a line of best fit is a curve which can be predictable or …… .

2 Nasma found a newspaper article about nanoscience. Nanoparticles are used for many things including perfumes.

   There was increasing evidence that inhaled nanoparticles could cause lung inflammation. [quote from Professor Ken Donaldson]

   Discuss the type of experiment and the data you would expect to see to support this conclusion.

NEXT TIME YOU…

. . . read scientific claims, think carefully about the evidence that should be there to back up the claim.

DID YOU KNOW?

Newland's 'rule of octaves' was devised to account for the apparent fact that the elements fell into groups rather like the notes on the musical scale. Mendeleev was an addict to the game of patience and used the card game to develop the periodic table of elements.

KEY POINTS

1 Drawing lines of best fit help us to study the relationship between variables.

2 The possible relationships are linear, positive and negative; directly proportional; predictable and complex curves.

3 Conclusions must go no further than the data available.

4 The reliability and validity of data can be checked by looking at other similar work done by others, perhaps on the Internet. It can also be checked by using a different method or by others checking your method.

H8 Scientific evidence and society

"I WAS ABDUCTED BY ALIENS!" FULL STORY

WE WILL ALL BE DESTROYED BY GREY GOO!

Nanotechnology could lead to better fuel cells, new materials and better ways of dealing with health problems, but we must also know what the dangers are.

Prince Charles said that nanotechnology was a triumph of human ingenuity, He also said that we must pay proper attention to the risks.

Continued page 12

Now you have reached a conclusion about a piece of scientific research. So what is next? If it is pure research then your fellow scientists will want to look at it very carefully. If it affects the lives of ordinary people then society will also want to examine it closely.

You can help your cause by giving a balanced account of what you have found out. It is much the same as any argument you might have. If you make ridiculous claims then nobody will believe anything you have to say.

Be open and honest. If you only tell part of the story then someone will want to know why! Equally, if somebody is only telling you part of the truth, you cannot be confident with anything they say.

a) An advert for a breakfast cereal claims that it has 'extra folic acid'. What information is missing? Is it important?

You must be on the lookout for people who might be biased when representing scientific evidence. Some scientists are paid by companies to do research. When you are told that a certain product is harmless, just check out who is telling you.

b) Bottles of perfume spray contain this advice 'This finished product has not been tested on animals'. Why might you mistrust this statement.

Suppose you wanted to know about the pollution effects of burning waste in a local incinerator. Would you ask the scientist working for the incinerator company or one working in the local university?

We also have to be very careful in reaching judgements according to who is presenting scientific evidence to us. For example, if the evidence might provoke public or political problems, then it might be played down.

Equally others might want to exaggerate the findings. They might make more of the results than the evidence suggests. Take as an example the quarrying of limestone. Local environmentalists may well present the same data in a totally different way to those with a wider view of the need for building materials.

c) Check out some websites on limestone quarrying in the National Parks. Get the opinions of the environmentalists and those of the quarrying companies. Try to identify any political bias there might be in their opinions.

The status of the experimenter may place more weight on evidence. Suppose a quarrying company wants to convince an enquiry that it is perfectly reasonable to site a quarry in remote moorland in the UK. The company will choose the most eminent scientist in that field who is likely to support them. The small local community might not be able to afford an eminent scientist. The enquiry needs to be very careful to make a balanced judgement.

SUMMARY QUESTIONS

1 Copy and complete using the words below:

  status   balanced   bias   political

  Evidence from scientific investigations should be given in a …… way. It must be checked for any …… from the experimenter.
  Evidence can be given too little or too much weight if it is of …… significance.
  The …… of the experimenter is likely to influence people in their judgement of the evidence.

2 Collect some newspaper articles to show how scientific evidence is used. Discuss in groups whether these articles are honest and fair representations of the science. Consider whether they carry any bias.

3 Petcoke is a high carbon product from refined oil. It can be used in power stations and cement works. Owners of the Drax power station, which is running a trial use of the fuel, claim that it is cheaper than coal and can be used without harmful effects. Other groups claim that it is 'dirty fuel' and will cause environmental and health problems. Suppose you were living near Drax power station. Who would you trust to tell you if petcoke was a safe fuel?

KEY POINTS

1 Scientific evidence must be presented in a balanced way that points out clearly how reliable and valid the evidence is.
2 The evidence must not contain any bias from the experimenter.
3 The evidence must be checked to appreciate whether there has been any political influence.
4 The status of the experimenter can influence the weight placed on the evidence.

H9

How is science used for everybody's benefit?

LEARNING OBJECTIVES

1 How does science link to technology?
2 How is science used and abused?
3 How are decisions made about science?
4 What are the limitations of science?

Coal has been burned for centuries as a source of energy. Experiments to capture the energy from burning coal were being carried out in the early 18th century. Thomas Newcomen's experimentation led to him inventing a steam engine.

James Watt was asked to repair an engine that used steam and realised that it would only go a few strokes at a time. He experimented still further to increase the efficiency of the steam engine. The unit of 'power' was named in his honour.

These investigations and their development into technologies were the foundation of the industrial revolution. At this time it seemed to ordinary people that the only problems with coal were the smell and the soot. The people who were making the most money out of these new developments could live well away from the pollution.

Coal remains a very important fuel – 38% of electricity is generated from coal. World-wide coal consumption is expected to grow by 40% before 2025.

Investigations into the effect of carbon dioxide on the atmosphere have shown that increased amounts of carbon dioxide are leading to global warming. Burning coal produces 9 billion tonnes of CO_2 each year, about a third of all CO_2 emissions world-wide.

Investigations have shown that new ways of burning coal, e.g. fluidised bed, mean that less CO_2 per kWh is produced. However CO_2 is still produced. It is too expensive to separate from the waste gases and so still contributes to global warming.

There are two approaches to cutting down CO_2 production. We can burn less fossil fuel or we can remove carbon dioxide from the waste gases of power stations and motor vehicles.

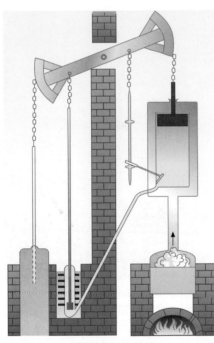

Figure 1 Newcomen's engine

DID YOU KNOW?

In 1712 Thomas Newcomen built his first engine on top of a water-filled mine shaft and used it to pump water out of the mine. Wonder where he got the fuel from . . . ?

Figure 2 A smoky steam engine

Figure 3 A coal-fired power station

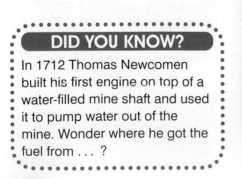

Experiments have taken place to change coal into CO_2 and H_2. The carbon from the coal is reacted with hot water to produce CO and H_2, then the CO and H_2O gives CO_2 and H_2. The hydrogen has five times the energy per kilogram as coal. It could be used to drive cars as well as power stations. Burning hydrogen in oxygen produces water.

The waste CO_2 is separated from the hydrogen. The CO_2 can then be piped into underground caverns left by oil and gas extraction. Here it cannot cause global warming – until it escapes!

The world's first large-scale CO_2 storage technology is being used in Norway. It is in an old gas field called Sleipner, and can store 600 billion tonnes of CO_2.

This technology was developed for removing CO_2 from natural gas using hot potassium carbonate. This process is still very expensive.

Experimentation suggests that the overall efficiency will be 60%.

There are many questions still to be answered. Can this technology be developed to allow all countries to be able to afford it?

There are questions which science cannot answer. For example, 'Should we develop technology or reduce energy use to reduce the effects of global warming?' Questions like this will have to be answered by the politicians that we elect.

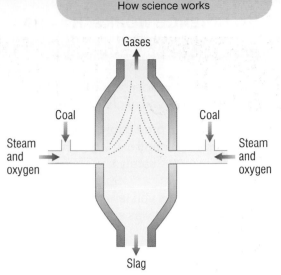

Figure 4 Changing coal into CO_2 and H_2

Figure 5 Sleipner field oil rig

KEY POINTS

1 Scientific knowledge can be used to develop technologies.
2 People can exploit scientific and technological developments to suit their own purposes.
3 The uses of science and technology can raise ethical, social, economic and environmental issues.
4 These issues are decided upon by individuals and by society.
5 There are many questions left for science to answer. But science cannot answer questions that start with 'Should we .... ?'

SUMMARY QUESTIONS

Use the account of the development of coal technology to answer these questions:

1 What early scientific work enabled James Watt to improve on the steam engine?

2 Use your own knowledge and that in the account above to describe some of the different ways in which the energy from coal has been used.

3 a) Identify some of these issues raised by the use of coal as a fuel source:
   i) ethical, ii) social, iii) economic, iv) environmental.
   b) Which of these issues are decided by individuals and which by society?

1 a) Fit these words into order. They should be in the order that you might use them in an investigation.

design; prediction; conclusion; method; repeat; controls; graph; results; table; improve; safety

2 a) How would you tell the difference between an opinion that was scientific and a prejudiced opinion?

b) Suppose you were investigating the amount of gas produced in a reaction. Would you choose to investigate a categoric, continuous or ordered variable? Explain why.

c) Explain the difference between a causal link between two variables and one which is due to association.

3 You might have observed that marble statues weather badly where there is air pollution. You ask the question why. You use some accepted theory to try to answer the question.

a) Explain what you understand by a hypothesis.

b) Sulfur dioxide in the air forms acids that attack the statues. This is a hypothesis. Develop this into a prediction.

c) Explain why a prediction is more useful than a hypothesis.

d) Suppose you have tested your prediction and have some data. What might this do for your hypothesis?

e) Suppose the data does not support the hypothesis. What should you do to the theory that gave you the hypothesis?

4 a) What do you understand by a fair test?

b) Suppose you were carrying out an investigation into what effect diluting acid had on its pH. You would need to carry out a trial. Describe what a trial would tell you about how to plan your method.

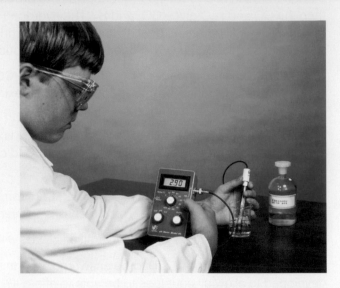

c) How could you decide if your results were reliable?

d) It is possible to calculate the effect of dilution on the pH of an acid. How could you use this to check on the accuracy of your results?

5 Suppose you were watching a friend carry out an investigation using the equipment shown on page 10. You have to mark your friend on how accurately he is making his measurements. Make a list of points that you would be looking for.

6 a) How do you decide on the range of a set of data?

b) How do you calculate the mean?

c) When should you use a bar chart?

d) When should you use a line graph?

7 a) What should happen to anomalous results?

b) What does a line of best fit allow you to do?

c) When making a conclusion, what must you take into consideration?

d) How can you check on the reliability of your results?

8 a) Why is it important when reporting science to 'tell the truth, the whole truth and nothing but the truth'?

b) Why might some people be tempted not to be completely fair when reporting their opinions on scientific data?

9 a) 'Science can advance technology and technology can advance science.' What do you think is meant by this statement?

b) Who answers the questions that start with 'Should we . . . '?

10 Glass has always been used for windows in buildings and is increasingly being used for structural parts as well. It is important therefore to be able to find out the strength of glass. One measure of this is the force that can be applied to glass before it breaks. Glass bends under pressure. Laminated glass is in three layers. Glass on the outside sandwiching a polymer layer. This strengthens the glass.

An experiment was carried out to find out how far laminated glass would bend. The glass was supported on two wooden blocks and a load line drawn half-way between the blocks. A load was then placed on this load line and the amount of bend in the glass was measured. The load was gradually increased. Another plate of glass was then used for a second set of results.

Line load (kN/m) with 2–sided support

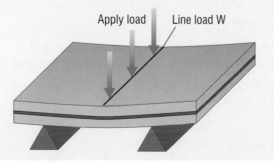

The results of the investigation are in the table.

Line load added (kN/m)	Bending (mm)	Bending (mm)
1	18	20
2	37	39
3	55	57
4	74	76
5	92	98

a) What was the prediction for this test?

b) What was the independent variable?

c) What was the dependent variable?

d) Suggest a control variable that should have been used.

e) Is there any evidence of a systematic error in this investigation? Explain your answer.

f) Is there any evidence for a random error? Explain your answer.

g) How could the investigation have its accuracy improved?

h) Was the precision of the bending measurement satisfactory? Provide some evidence for your answer from the data in the table.

i) What is the mean for the results at a line load of 5 kN/m?

j) Draw a graph of the results for the first test.

k) Draw a line of best fit.

l) Describe the pattern in these results.

m) What conclusion can you make?

n) How might you develop this technique to show the effect of the thickness of the polymer on the breaking point of the glass?

o) How might this information be used by architects wanting to protect buildings.

C1a | Products from rocks

Rocks, such as limestone from this quarry, provide us with many useful materials

What you already know

Here is a quick reminder of previous work that you will find useful in this unit:

- The chemical elements consist of atoms, which we represent by symbols.

- We can arrange the chemical elements in the periodic table.

- The substances in a mixture are not joined together chemically, so we can separate them again.

- We can represent chemicals using chemical formulae.

- We can summarise what happens in a chemical reaction using a word equation.

- The mass of the products formed in a chemical reaction is the same as the mass of the reactants they were formed from.

- When we burn fossil fuels we may produce new substances that can affect the environment.

- We can place the metals in a reactivity series. Metals higher up the series can displace metals lower down the series.

RECAP QUESTIONS

1 How many different types of atom are there in a jar of pure sulfur?

2 What are the symbols for the following elements: iron, oxygen, copper, sodium, chlorine, aluminium, calcium?

3 How could you separate a mixture of sand and salt?

4 Iron and chlorine react together to make iron chloride. Write a word equation for this reaction.

5 a) Describe in words what happens in this reaction:

   magnesium + oxygen → magnesium oxide

   b) In the reaction between magnesium and oxygen, the mass of magnesium and oxygen was 2.5 grams. How much magnesium oxide was formed?

6 When fossil fuels burn in plenty of air, what new substances are produced?

7 Magnesium ribbon is put into blue copper sulphate solution. The solution becomes colourless and a pinkish metal is produced.

   Then copper metal is put into a solution of silver nitrate. The solution turns blue and silver metal is produced.

   Arrange the three metals, copper, silver and magnesium, in order of reactivity. Put the most reactive first.

Making connections

My job takes me all round the world! I enjoy all the travelling — I always have. When I'm not flying for work I fly for pleasure — I've got a little light aircraft that I keep near my home.

This new airport building really will be a fantastic place for international passengers to arrive. It couldn't have been designed ten years ago — we just didn't have the materials that could have been used to build it then.

We have to keep all the runways in excellent condition for safety reasons. The runways are made of concrete, and with aeroplanes getting heavier all the time we have to be on the lookout for cracks. There's lots of research going on to work out the best way to make runways last as long as possible.

I have to drive about forty miles to work each day. The drive doesn't worry me, but the cost does. So I bought a diesel car which uses much less fuel. It still costs a lot, but it's a price I'm prepared to pay for living somewhere nice.

We have to make sure that the aircraft have all the fuel they need — it is no good if they can't take off because we haven't finished our job! A big airliner needs thousands of kilograms of fuel on board — they burn about six tonnes an hour, you know!

We've only flown once before, and we're very nervous. Our son and his wife emigrated to Australia last year, so this is the only way we can get to see them. Flying is exciting though, isn't it?!

It's our job to keep the aircraft well-maintained. Modern materials — especially alloys — make aeroplanes lighter, stronger and safer than ever before.

ACTIVITY

An airport shows how important science has become for the world we live in. Building materials, engineering materials for modern aircraft, and fuels – each of these relies on up-to-date scientific knowledge.

Work in groups to plan an exhibition to go on display for the passengers and travellers passing through an airport. You should design your exhibition to show how air transport relies on materials made by scientists. It should show the importance of these materials from the time people leave home to the time the aircraft gets off the ground.

You'll need to think about the environmental impact of airports and flying too. How will you get the message across that too much flying is bad for the planet without upsetting your airport sponsors? Use pictures, words, computer presentations – anything to make your message clear.

Flying is supposed to be so glamorous, but I fly so often it's just a bore. I mean, one first class lounge is just like another, isn't it?

Chapters in this unit

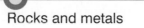

Rocks and building   Rocks and metals   Crude oil

C1a 1.1 Atoms, elements and compounds

LEARNING OBJECTIVES

1 What are elements made of?
2 How do we represent atoms and elements?
3 What are atoms and how do they join together?

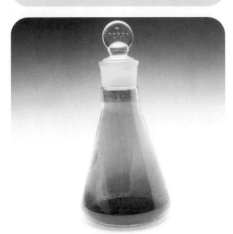

Figure 1 An element contains only ***one*** type of atom – in this case bromine

Look at the things around you and the substances that they are made from – wood, metal, plastic, glass . . . the list is almost endless. Look further and the number of different substances is mind-boggling.

All substances are made of **atoms**. We think that there are about 100 different types of atom. These can combine in a huge variety of ways to give us all those different substances.

Some substances are made up of only one type of atom. We call these substances **elements**. As there are only about 100 different types of atom, there are only about 100 different elements.

a) How many different types of atom are there?
b) Why can you make millions of different substances from these different types of atom?

Elements can have very different properties. For example, elements such as silver, copper and gold are shiny solids. Other elements such as oxygen, nitrogen and chlorine are gases.

Atoms have their own symbols

The name that an element has depends on the language. For example, sulfur is called *Schwefel* in German and *azufre* in Spanish! Because a lot of scientific work is international, it is important that we have symbols for elements that everyone understands. You can see these symbols in the periodic table in Figure 2.

Group numbers

1	2											3	4	5	6	7	0
			H 1 Hydrogen														He 2 Helium
Li 3 Lithium	Be 4 Beryllium											B 5 Boron	C 6 Carbon	N 7 Nitrogen	O 8 Oxygen	F 9 Fluorine	Ne 10 Neon
Na 11 Sodium	Mg 12 Magnesium											Al 13 Aluminium	Si 14 Silicon	P 15 Phosphorus	S 16 Sulfur	Cl 17 Chlorine	Ar 18 Argon
K 19 Potassium	Ca 20 Calcium	Sc 21 Scandium	Ti 22 Titanium	V 23 Vanadium	Cr 24 Chromium	Mn 25 Manganese	Fe 26 Iron	Co 27 Cobalt	Ni 28 Nickel	Cu 29 Copper	Zn 30 Zinc	Ga 31 Gallium	Ge 32 Germanium	As 33 Arsenic	Se 34 Selenium	Br 35 Bromine	Kr 36 Krypton
Rb 37 Rubidium	Sr 38 Strontium	Y 39 Yttrium	Zr 40 Zirconium	Nb 41 Niobium	Mo 42 Molybdenum	Tc 43 Technetium	Ru 44 Ruthenium	Rh 45 Rhodium	Pd 46 Palladium	Ag 47 Silver	Cd 48 Cadmium	In 49 Indium	Sn 50 Tin	Sb 51 Antimony	Te 52 Tellurium	I 53 Iodine	Xe 54 Xenon
Cs 55 Caesium	Ba 56 Barium	Lanthanum see below	Hf 72 Hafnium	Ta 73 Tantalum	W 74 Tungsten	Re 75 Rhenium	Os 76 Osmium	Ir 77 Iridium	Pt 78 Platinum	Au 79 Gold	Hg 80 Mercury	Tl 81 Thallium	Pb 82 Lead	Bi 83 Bismuth	Po 84 Polonium	At 85 Astatine	Rn 86 Radon
Fr 87 Francium	Ra 88 Radium	Actinium see below															
↑ The alkali metals	↑ The alkaline earth metals					↑ The transition metals										↑ The halogens	↑ The noble gases

Lanthanides →

La 57 Lanthanium	Ce 58 Cerium	Pr 59 Praseodymium	Nd 60 Neodymium	Pm 61 Promethium	Sm 62 Samarium	Eu 63 Europium	Gd 64 Gadolinium	Tb 65 Terbium	Dy 66 Dysprosium	Ho 67 Holmium	Er 68 Erbium	Tm 69 Thulium	Yb 70 Ytterbium	Lu 71 Lutetium

Actinides →

Ac 89 Actinium	Th 90 Thorium	Pa 91 Protactinium	U 92 Uranium	Np 93 Neptunium	Pu 94 Plutonium	Am 95 Americium	Cm 96 Curium	Bk 97 Berkelium	Cf 98 Californium	Es 99 Einsteinium	Fm 100 Fermium	Md 101 Mendelevium	No 102 Nobelium	Lr 103 Lawrencium

Figure 2 The periodic table shows the symbols for the elements and a lot of other information too

The symbols in the periodic table represent atoms. For example, O represents an atom of oxygen while Na represents an atom of sodium. The elements in the table are arranged in vertical columns, called *groups*. Each group contains elements with similar chemical properties.

c) Why do we need symbols to represent atoms of different elements?

d) A particular element in the periodic table is a reactive metal. What properties are the elements in the same column as this element likely to have?

Atoms, elements and compounds

Almost all of the substances around us are not pure elements. Most substances are made up of different atoms joined together to form **compounds**. Some compounds are made from just two types of atom joined together (e.g. water, made from hydrogen and oxygen). Other compounds consist of many different types of atom.

Atoms are made up of a tiny central *nucleus* with *electrons* around it.

Look at the diagram in Figure 3.

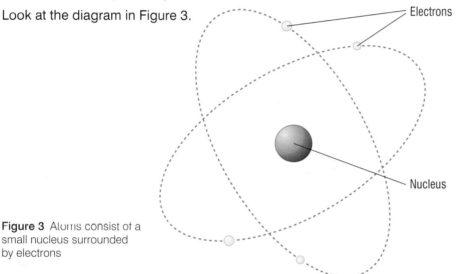

Electrons

Nucleus

Figure 3 Atoms consist of a small nucleus surrounded by electrons

Sometimes atoms react together by transferring electrons to form chemical bonds. Other atoms may share electrons to form chemical bonds. No matter how they are formed, chemical bonds hold atoms tightly to each other.

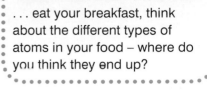
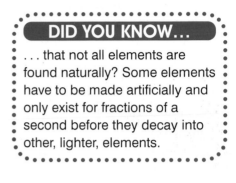

DID YOU KNOW...

. . . that not all elements are found naturally? Some elements have to be made artificially and only exist for fractions of a second before they decay into other, lighter, elements.

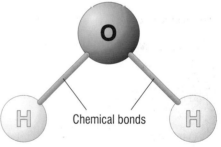

O

H   Chemical bonds   H

Figure 4 Two or more atoms bonded together are called a **molecule**. Look at the chemical bonds holding the hydrogen and oxygen atoms together in the water molecule.

SUMMARY QUESTIONS

1 Copy and complete using the words below:

    atoms   **bonds**   **molecule**   **sharing**

All elements are made up of .... When two or more atoms join together a ...... is formed. The atoms in elements and compounds are held together by giving, taking or ...... electrons. We say that they are held tightly together by chemical ....

2 Explain the following statement: 'When two elements are mixed together they can often be separated quite easily, but when two elements are combined in a compound they can be very difficult to separate.'

3 a) Draw diagrams to help explain the difference between an element and a compound, using actual examples.

  b) Draw a labelled diagram to show the structure of an atom.

KEY POINTS

1 All substances are made up of atoms.
2 Elements contain only one type of atom.
3 Different atoms can bond together by giving, taking or sharing electrons, to form compounds.

C1a 1.2 Limestone and its uses

LEARNING OBJECTIVES

1 What are the uses of limestone?
2 What happens when we heat limestone?

Figure 1 These white cliffs are made of chalk. This is one type of limestone, formed from the shells of tiny sea plants.

Figure 2 St Paul's Cathedral in London is built from limestone blocks

Uses of limestone

Limestone is a rock that is made mainly of **calcium carbonate**. Some types of limestone were formed from the remains of tiny animals and plants that lived in the sea millions of years ago. We dig limestone out of the ground in quarries all around the world. It has many important uses, including use as a building material.

Many important buildings around the world are made of limestone. We can cut and shape the stone taken from the ground into blocks. These can be placed one on top of the other, like bricks in a wall. We have used limestone in this way to make buildings for hundreds of years.

We can also process limestone to make other building materials. Powdered limestone can be heated to high temperatures with a mixture of sand and sodium carbonate (soda) to make **glass**.

Powdered limestone can also be heated with powdered clay to make **cement**. When we mix cement powder with water, sand and crushed rock, a slow chemical reaction takes place. The reaction produces a hard, stone-like building material called **concrete**.

a) What is limestone made of?
b) How do we use limestone to make buildings?

Figure 3 This building contains plenty of concrete which is made from limestone

Figure 4 Glass gives us buildings that let in natural light and which protect us from the weather

Heating limestone

The chemical formula for calcium carbonate is $CaCO_3$. This tells us that for every calcium atom (Ca) there is one carbon atom (C) and three oxygen atoms (O).

When we heat limestone strongly it breaks down to form a substance called **quicklime** (calcium oxide). Carbon dioxide is also produced in this reaction. Breaking down a chemical by heating is called **thermal decomposition**.

To make lots of quicklime this reaction is done in a furnace called a *lime kiln*. We fill the kiln with crushed limestone and heat it strongly using a supply of hot air. Quicklime comes out of the bottom of the kiln, while waste gases leave the kiln at the top.

We can show the thermal decomposition reaction by a word equation:

$$\text{calcium carbonate} \xrightarrow{\text{heat}} \text{calcium oxide} + \text{carbon dioxide}$$
$$\text{(quicklime)}$$

A rotary lime kiln

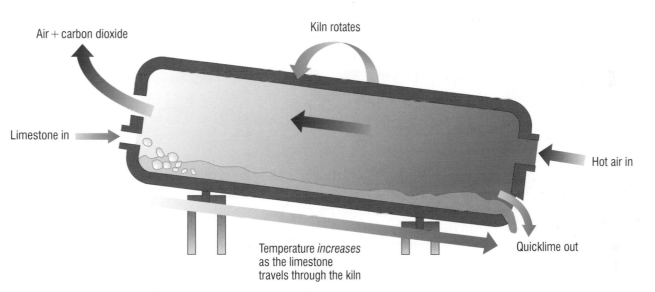

Figure 5 Quicklime is produced in a lime kiln. It is often produced in a **rotary kiln**, where the limestone is heated in a rotating drum. This makes sure that the limestone is thoroughly mixed with the stream of hot air so that it decomposes completely.

SUMMARY QUESTIONS

1 Copy and complete using the words below:

   $CaCO_3$ calcium   cement   concrete   glass

   Limestone is mostly made of …… carbonate (whose chemical formula is ……). As well as being used to produce blocks of building material, limestone can be used to produce …… , …… and …… that can also be used in building.

2 Produce a poster to show how limestone is used in building.

3 The stone roof of a building is supported by columns made of limestone. Why might this be unsafe after a fire in the building? Explain the chemical process involved in weakening the structure.

KEY POINTS

1 Limestone is made mainly of calcium carbonate.
2 Limestone is widely used in building.
3 Limestone breaks down when we heat it strongly (thermal decomposition) to make quicklime and carbon dioxide.

C1a 1.3 Decomposing carbonates

LEARNING OBJECTIVES

1 Do other carbonates behave in the same way as limestone?
2 What happens to the atoms in a chemical reaction?
3 Is the mass of the products in a chemical reaction related to the mass of the reactants?

PRACTICAL

Investigating carbonates
You can investigate the thermal decomposition of carbonates by heating samples in a Bunsen flame. Imagine that you have been given samples of the carbonates listed below. What kinds of changes might tell you if a sample decomposes when you heat it?

Figure 1 Investigating the thermal decomposition of a solid

Carbonate samples:
sodium carbonate,
potassium carbonate,
magnesium carbonate,
zinc carbonate,
copper carbonate.

In the last spread we saw that limestone consists mainly of calcium carbonate. This decomposes when we heat it, producing quicklime (calcium oxide) and carbon dioxide. Calcium is an element in Group 2 of the periodic table (see periodic table on page 24). As we have already seen, the elements in a group tend to behave in the same way – so does magnesium carbonate also decompose when you heat it? And what about other carbonates too?

a) Why might you expect magnesium carbonate to behave in a similar way to calcium carbonate?

In an investigation into the behaviour of carbonates, a student draws the following conclusions when he heats samples of carbonates:

Calcium carbonate	Sodium carbonate	Potassium carbonate	Magnesium carbonate	Zinc carbonate	Copper carbonate
✓	✗	✗	✓	✓	✓

(✓ = decomposes, ✗ = does not decompose)

b) To which group in the periodic table do sodium and potassium belong? (See page 24.)
c) To which group in the periodic table do magnesium and calcium belong?
d) What do these conclusions suggest about the behaviour of the carbonates of elements in Group 1 and Group 2?
e) Can you be certain about your answer to question d)? Give reasons.

Investigations like this show that when many carbonates are heated in a Bunsen flame they decompose. They form the metal oxide and carbon dioxide – just as calcium carbonate does. Sodium and potassium carbonate do not decompose at the temperature in the Bunsen flame.

Balancing equations

We can represent the decomposition of carbonates using chemical equations. These use chemical formulae for elements and compounds, and help us to see how much of each chemical is reacting. Representing reactions in this way is better than using word equations, for three reasons:

● word equations are only useful if everyone who needs to read them speaks the same language,
● word equations do not tell us how much of each substance is involved in the reaction,
● word equations can get very complicated when lots of chemicals are involved.

When calcium carbonate decomposes, we can show the chemical reaction like this:

$$CaCO_3 \rightarrow CaO + CO_2$$

This equation is **balanced** – there is the same number of each type of atom on both sides of the equation. This is very important, because atoms cannot be created or destroyed in a chemical reaction. This also means that the mass of the products formed in the reaction is equal to the mass of the reactants.

We can check if an equation is balanced by counting the number of each type of atom on either side of the equation. If the numbers are equal, then the equation is balanced.

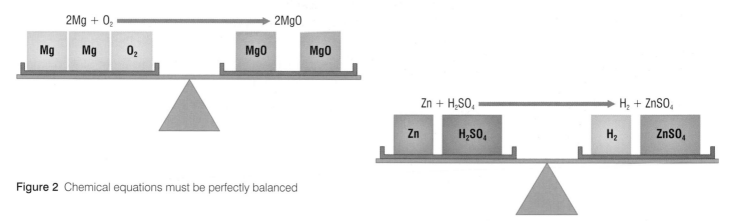

Figure 2 Chemical equations must be perfectly balanced

Magnesium carbonate decomposes like this:

$$MgCO_3 \rightarrow MgO + CO_2$$

Notice how the symbol for magnesium (Mg) has simply replaced the symbol for calcium (Ca). Everything else has stayed the same.

f) Zinc carbonate has a similar formula to calcium carbonate and decomposes in the same way. Write the balanced equation for this reaction.

In some cases it is not so easy to balance an equation. Here is an example of a reaction which shows how it must be written to produce a balanced equation.

When magnesium reacts with hydrochloric acid, hydrogen gas is given off. Magnesium chloride is also formed. Magnesium chloride has the formula $MgCl_2$ and hydrogen gas is H_2 so the equation starts off as:

$$Mg + HCl \rightarrow MgCl_2 + H_2$$

But this means that there are not enough H and Cl atoms on the left-hand side. By adding another HCl molecule we can balance the equation, which now reads:

$$Mg + 2HCl \rightarrow MgCl_2 + H_2$$

To check that this equation is balanced, we add up the different atoms on either side. There are two magnesium atoms, two hydrogen atoms and two chlorine atoms on each side of the equation – so it is balanced.

SUMMARY QUESTIONS

1 Give the general word equation for the thermal decomposition of a metal carbonate.

2 a) A mass of 2.5 g of barium carbonate ($BaCO_3$) completely decomposes when it is heated. What is the total mass of products formed in this reaction?

b) Write a balanced equation to show the reaction in a).

KEY POINTS

1 Some carbonates decompose when we heat them in a Bunsen flame.

2 There is the same number of each type of atom on each side of a balanced chemical equation.

3 The mass of reactants is the same as the mass of the products in a chemical reaction.

C1a 1.4   Quicklime and slaked lime

Limestone is used very widely as a building material. We can also process it to make other materials used for building.

As we saw on page 27, quicklime is produced when we heat limestone strongly. The calcium carbonate in the limestone undergoes thermal decomposition.

When we add water to quicklime it reacts to produce **slaked lime**. Slaked lime's chemical name is **_calcium hydroxide_**. This reaction gives out a lot of heat.

$$\text{calcium oxide} + \text{water} \rightarrow \text{calcium hydroxide}$$
$$CaO + H_2O \rightarrow Ca(OH)_2$$

Although it is not very soluble, we can dissolve a little calcium hydroxide in water. After filtering, this produces a colourless solution called **_lime water_**. We can use lime water to test for carbon dioxide. When carbon dioxide is bubbled through clear lime water the solution turns cloudy. This is because calcium carbonate is formed, which is insoluble in water.

$$\text{calcium hydroxide} + \text{carbon dioxide} \rightarrow \text{calcium carbonate} + \text{water}$$
$$Ca(OH)_2 + CO_2 \rightarrow CaCO_3 + H_2O$$

a) What substance do we get when quicklime reacts with water?
b) Describe how we can make lime water from this substance.

You can explore the reactions of limestone using some very simple apparatus.

PRACTICAL

Investigating the chemical reactions of limestone

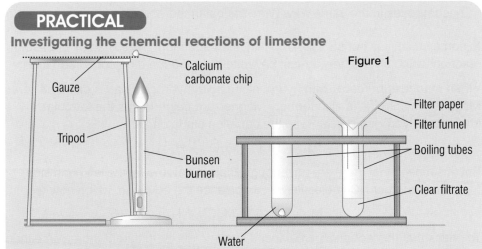

Figure 1

Heat the calcium carbonate chip very strongly, making it glow. This produces calcium oxide (quicklime). Let the calcium oxide cool down.
Then you react the calcium oxide with a few drops of water to produce calcium hydroxide (slaked lime).
When you dissolve this in water and filter, it produces lime water.
Carbon dioxide bubbled through the lime water produces calcium carbonate and the solution goes cloudy.

● The reaction between quicklime and water gives out a lot of energy. What do you observe during the reaction?
● Why does bubbling carbon dioxide through lime water make the solution go cloudy?

Mortar

About 6000 years ago the Egyptians heated limestone strongly in a fire and then combined it with water. This produced a material that hardened with age. They used this material to plaster the pyramids. Nearly 4000 years later, the Romans mixed slaked lime with sand and water to produce **mortar**.

Mortar holds other building materials together – for example, stone blocks or bricks. It works because the lime in the mortar reacts with carbon dioxide in the air, producing calcium carbonate again. This means that the bricks or stone blocks are effectively held together by stone, which makes the construction very strong.

$$\text{slaked lime} + \text{carbon dioxide} \rightarrow \text{calcium carbonate} + \text{water}$$
$$Ca(OH)_2 \quad + \quad CO_2 \quad \rightarrow \quad CaCO_3 \quad + \quad H_2O$$

The amount of sand in the mixture is very important – too little sand and the mortar shrinks as it dries. Too much sand makes it too weak.

Even today, mortar is still used widely as a building material. However, modern mortars can be used in a much wider range of ways than in ancient Egyptian and Roman times.

Figure 2 The original lime mortar has flaked away from the surface of the Sphinx in Egypt, and many of the stones are now missing

Figure 3 Lime mortar should be used to repair old buildings

SUMMARY QUESTIONS

1 Copy and complete using the words below:

> **calcium hydroxide**   **carbonate**   **carbon dioxide**   **mortar**
> **quicklime**   **thousands**

Limestone has been used in building for ...... of years. When it is heated, limestone produces ......, also called calcium oxide. If calcium oxide is reacted with water .... is produced. This can be combined with sand to produce ......, used to hold building materials like bricks together. Mortar relies on .... in the air to produce a chemical reaction in which calcium ...... is formed.

2 a) When quicklime reacts with water, slaked lime is produced. Write a balanced equation to show the reaction.
   b) The slaked lime in mortar reacts with carbon dioxide in the air to produce calcium carbonate and water. Write a balanced equation for this reaction.

KEY POINTS

1 When water is added to quicklime it produces slaked lime.
2 Lime mortar is made by mixing slaked lime with sand and adding water.

C1a 1.5   Cement, concrete and glass

Figure 1 Lime mortar is not suitable for building pools since it will not harden when in contact with water

Cement

Although lime mortar holds bricks and stone together very strongly it does not work in all situations. In particular, lime mortar does not harden very quickly. It will not set at all where water prevents it from reacting with carbon dioxide.

The Romans realised that they needed to add something to lime mortar to make it set in wet conditions. They found that adding brick dust or volcanic ash to the mortar mixture enabled the mortar to harden even under water. This method remained in use until the 18th century.

Then people found that heating limestone with clay in a kiln produced **cement**. Much experimenting led to the invention of *Portland cement*. We make this from a mixture of limestone, clay and other minerals which are heated and then ground into a fine powder.

This type of cement is still in use today. The mortar used to build a modern house consists of a mixture of Portland cement and sand. This sets when it is mixed thoroughly with water and left for a few days.

Figure 2 Portland cement was invented nearly 200 years ago. It is still in use all around the world today.

a) What does lime mortar need in order to set hard?
b) Why will lime mortar not set under water?

Concrete

Sometimes builders add stones or crushed rocks to the mixture of water, cement and sand. When this sets, it forms a hard, stone-like building material called **concrete**.

This material is very strong. It is especially good at resisting forces which tend to squeeze or crush it. We can make concrete even stronger by pouring it around steel rods or bars and then allowing it to set. This makes *reinforced concrete*, which is also good at resisting forces that tend to pull it apart.

Figure 3 Badly designed reinforcement caused the concrete roof of this terminal building at Charles de Gaulle Airport in France to collapse in May 2004

PRACTICAL

Which mixture makes the strongest concrete?

Try mixing different proportions of cement, gravel and sand, then adding water, to find out how to make the strongest concrete.

- How did you test the concrete's strength?
- How could you improve the reliability of your data?

Figure 4 Glass can produce some spectacular buildings

Glass

We can also use limestone to make a very different kind of building material. When powdered limestone is mixed with sand and sodium carbonate and then heated strongly it produces **glass**.

Glass is very important in buildings since it allows us to make them both weatherproof and light. Hundreds of years ago only very rich people could afford glass for their windows. So ordinary people's buildings must have been very cold and very dark during the winter!

Modern chemists have developed glass with many different properties. This makes it possible to design buildings that would have been impossible to build even 50 years ago.

GET IT RIGHT!

You don't need to know the details of the industrial processes for making glass – just how it is made.

SUMMARY QUESTIONS

1 Write down three ways that limestone is used in building other than as limestone blocks.

2 List the different ways in which limestone has been used to build your home or school.

3 Glass has been advertised as 'insulation you can see through!' Explain what the advertiser meant by this.

4 Concrete and glass are commonly used building materials. Evaluate the use of:
   a) concrete to make a path rather than using bricks
   b) glass to make a window pane rather than using perspex.

KEY POINTS

1 Cement is made by heating limestone with clay in a kiln.
2 Concrete is made by mixing crushed rocks, cement and sand with water.
3 Glass is made by heating powdered limestone, sand and sodium carbonate together very strongly.

C1a 1.6   Building materials – from old to new

Out of the past . . .

Since ancient times people have always needed somewhere to shelter. In hot countries people need to find somewhere cool during the day. In cold countries they need somewhere warm – and in wet, windy Britain we often need somewhere to get out of the rain!

In the past, people often had to move around since they did not work in one place. People who looked after animals or who worked in different places at different times of the year built shelters. They used wood and any other natural materials that they could find at that place. They built very simple homes that could be put up quickly. They did not use large amounts of material that had to be carried over long distances.

Once people started to settle down and live in one place, it was worth building a house which was bigger and which took longer to build. People began to use materials like stone, and to develop new materials like bricks and concrete. It was more difficult to build well with these new materials. This meant it was often necessary for people with special skills to be involved in building.

Figure 1 A charcoal burner's hut. The charcoal burner had to watch the kiln constantly, so it was necessary to live right next to it. Charcoal was sold as a fuel. It burned with an intense heat, much hotter than wood.

QUESTIONS

1 Why did ancient people use local materials for building rather than materials that had to be brought from a long distance away?

2 Why did people need special skills to work with materials like bricks and mortar?

. . . and into the future

As new building materials were developed, new ways of using them were found. Buildings could be made bigger and taller, and new designs were possible. People could live in big houses, grand houses, small houses – and even flats and bungalows.

As our understanding of the chemistry of materials grew, artificial materials like plastics and metals began to replace natural materials like wood. Because the properties of these new materials can be controlled, they can be made to order. They are produced to have exactly the right strength for the job they have to do. As materials scientists continue to produce exciting new materials, who knows what homes in the future may look like?!

ACTIVITY

a) Find some pictures of old houses. Find out what they are made of and think about why people chose these materials.

b) Design a house of the future. Clearly label the materials you use and their properties. Be creative – you can use materials that haven't been invented yet!

*life*style

Ancient or Modern?

Anna and Simon are buying their second house. It's a modern one, on a big estate. They're hoping to make it look a bit different, to reflect their personalities and lifestyle.

" We'd really like an old house, but there's nothing like that around here. So we want to use old building materials to make the house look older than it really is. We think that we might use natural limestone to cover up the bricks, and we're going to put an old wooden door in to replace the PVC front door. We might even take out the modern windows too. We want it to look more like a cottage than a house on a big estate. "

Roger and Su moved into their house a while ago. It's old, and needs some work done on it. Some of the windows leak, and there's a lot of painting to do! They want to try to make the house easy to maintain.

" This house is very old, and it really needs updating. With our children growing up, we really don't have time to look after the house. If we put modern doors and windows in, we won't have to worry about painting them. The windows will be double-glazed too, so the house will be much warmer in winter. And that old metal gutter round the roof – we'll replace that with white PVC plastic. So much more modern! "

ACTIVITY

Imagine that you are the host of a TV makeover show. Both couples in the magazine are guests on the show. They need your advice about how they should makeover their houses. Write and present a short script for the programme, presenting your advice to each couple. Remember that you'll not only need to tell them what they should do but why they should do it.

ACTIVITY

Decisions, decisions

There's a huge choice of building materials available for people who want to update their homes – modern scientists have seen to that!

Discuss the issues below:
- Should people just 'do want they fancy' when they decide to update a house?
- Are Anna and Simon right to think about making their modern house look old? Why?
- Should Roger and Su try to make their house as easy as possible to maintain? What advice would you give them?

SUMMARY QUESTIONS

1 a) Jim has a sample of a pure element. How many different types of atom are there in it?

   b) What do we call the vertical columns of elements in the periodic table?

   c) Three elements are in the same column of the periodic table. What does this tell us about the reactions of these three elements?

   d) The atoms of two different elements react together to form a compound. Why is it difficult to separate the two elements when this has happened?

2 The diagram shows a design for a handwarmer which uses quicklime and water. To activate the handwarmer you squeeze the container. This breaks the capsule containing the water so that it mixes with the quicklime.

   a) Using complete sentences, describe how the handwarmer works.

   b) Write down a balanced chemical equation for the reaction in the handwarmer, using the correct symbols.

   c) Is the handwarmer re-usable or disposable? Give reasons for your answer.

3 a) A set of instructions for making concrete reads:

   'To make good, strong concrete, thoroughly mix together 4 buckets of gravel, 3 buckets of sand and one bucket of cement. When you have done this, add half a bucket of water.'

   Copy and complete the table showing the percentage of each ingredient in the concrete mixture. Give your answers to the nearest whole number.

Ingredient	gravel	sand	cement	water
Number of buckets				
Percentage				

   b) Describe an investigation you could try to see which particular mixture of gravel, sand and cement makes the strongest concrete. What would you vary, what would you keep the same and how would you test the 'strength' of the concrete?

EXAM-STYLE QUESTIONS

1 The diagram shows a molecule of ammonia, NH_3.

H—N—H
|
H

Match the words **A**, **B**, **C** and **D** with spaces **1** to **4** in the sentences.

A bonds

B electrons

C elements

D symbols

Ammonia is a compound made from two ....**1**......... .
The atoms in the molecule are represented by ....**2**......... .
The atoms in ammonia are held together by chemical ....**3**.......... .
Each atom has a nucleus surrounded by ....**4**.......... .

(4)

2 The diagram shows stages in making cement and concrete.

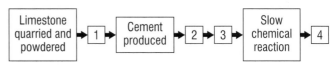

Match statements **A**, **B**, **C** and **D** with the numbers **1** to **4** to describe what happens in this process.

A cement mixed with sand and crushed rock

B concrete produced

C limestone heated in a kiln with clay

D water added to mixture (4)

3 (a) Slaked lime is made by reacting quicklime with:

   A carbon dioxide

   B oxygen

   C sulfuric acid

   D water (1)

   (b) The chemical name for slaked lime is:

   A calcium chloride

   B calcium hydroxide

   C calcium oxide

   D calcium sulfate (1)

   (c) Slaked lime can be used to make:

   A bricks

   B clay

   C mortar

   D quicklime (1)

(d) Lime water goes cloudy when reacted with carbon dioxide. Which substance is produced?

    A   calcium carbonate

    B   calcium chloride

    C   calcium oxide

    D   calcium sulfate   (1)

4 Glass is used in almost all buildings.

  (a) Suggest **two** properties of glass that make it useful in buildings.   (2)

  (b) Suggest and explain one disadvantage of using glass in buildings.   (2)

5 One of the largest limestone quarries in the United Kingdom is near the town of Buxton. It is in the Peak District National Park, an area popular with tourists.

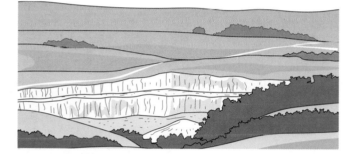

Suggest **three** social or environmental issues involved in quarrying limestone in the Peak District.   (3)

6 Mortars used in most modern buildings are made using cement.
A student tested the strength of a ready-mixed mortar. He did this by dropping a mass onto a small mortar beam from increasing heights until the beam broke in half. He tested 4 beams made from the mortar. His results were 20 cm, 50 cm, 65 cm and 15 cm.

  (a) What was the range of the student's results?   (2)

  (b) Work out the mean of his results.   (1)

  (c) Comment on the precision of his results.   (1)

  (d) (i) Besides cement, what was the other solid in the ready-mixed mortar?   (1)

     (ii) What other solid is needed to make concrete instead of mortar?   (1)

HOW SCIENCE WORKS QUESTIONS

Look at the standards for the testing of cement shown below and answer the questions that follow.

> **OFFICIAL DOCUMENT**
> **IMPORTANT**
>
> ### Standards for the testing of cement
>
> Cement must be tested to the following standards:
>
> - After 7 days use compressive test standard equipment to test the strength of the cement.
> - Three batches of already tested cement must also be treated in the same way.
> - The bowl must be wiped clean and 400 g of test cement added during 30 seconds. Add 400 g of the sand and mix for 120 seconds. Add 200 g of water and mix for 240 seconds.
> - The sand being used must be washed, heated in a kiln at 110°C and then passed through a sieve with holes of diameter 1 mm.
> - The mixing bowl used must be between 20 cm wide at the top, narrowing in a curve to 8 cm at the bottom and made of stainless steel.
> - The apparatus used must conform to that described in CTM19.

a) What was the dependent variable in this testing? (1)

b) What was the 'control group' in this testing? (1)

c) Why is it important to test cement that has already been tested? (1)

d) Why is it important to test the cement? (1)

e) Why is it important to give so much detail of how to test the cement? (1)

f) Who should NOT carry out these tests? Explain your answer. (2)

g) Who should carry out these tests? Explain your answer. (2)

C1a 2.1

Extracting metals

LEARNING OBJECTIVES

1 Where do metals come from?
2 How do we extract metals from the Earth?

Figure 1 The Angel of the North stands 20 metres tall, and is made of steel which contains a small amount of copper

Metals have been important to people for thousands of years. You can follow the course of history by the materials people used – from the Stone Age to the Bronze Age and then on to the Iron Age.

Where do metals come from?

Metals are found in the Earth's crust. We find most metals combined with other chemical elements, often with oxygen. This means that the metal must be chemically separated from its compounds before you can use it.

In some places there is enough of a metal or metal compound in a rock to make it worth extracting the metal. Then we call the rock a metal **ore**.

Whether it is worth extracting a particular metal depends on:

● how easy it is to extract it from its ore,
● how much metal the ore contains.

A few metals, such as gold and silver, are so unreactive that they are found in the Earth as the metals (elements) themselves. We say that they exist in their *native* state.

Sometimes a nugget of gold is so large it can simply be picked up. At other times tiny flakes have to be physically separated from sand and rocks by panning.

a) Where do we find metals in nature?
b) If there is enough metal in a rock to make it economic to extract it, what do we call the rock?
c) Why are silver and gold found as metals rather than combined with other elements?

Figure 2 Panning for gold. Mud and stones are washed away while the dense gold remains in the pan.

DID YOU KNOW...

... that gold in Wales is found in seams, just like coal – although not as thick, unfortunately! Gold jewellery was worn by early Welsh princes as a badge of rank. Welsh gold has been used in modern times to make the wedding rings of Royal brides.

☐ Oxygen	46%
☐ Silicon	28%
☐ Aluminium	8%
☐ Iron	5%
☐ Calcium	4%
☐ Sodium	3%
■ Magnesium	2%
■ Potassium	2%
■ Titanium	0.5%
☐ Hydrogen	0.5%
■ *All other elements*	1%

Figure 3 There are many different elements that go to make up the Earth's crust

How do we extract metals?

The way in which we extract a metal depends on its place in the *reactivity series*. The reactivity series lists the metals in order of their reactivity. The most reactive are placed at the top and the least reactive at the bottom.

A more reactive metal will displace a less reactive metal from its compounds. Carbon (a non-metal) will also displace less reactive metals from their oxides. We use carbon to extract metals from their ores commercially.

d) A metal cannot be extracted from its ore using carbon. Where is this metal in the reactivity series?

You can find many metals, such as copper, lead, iron and zinc, combined with oxygen. The compounds are called **metal oxides**. Because carbon is more reactive than each of these metals, you can use it to extract them from their ores.

When you heat the metal oxide with carbon, the carbon removes the oxygen from the metal oxide to form carbon dioxide. The reaction leaves the metal, as the element, behind:

$$\text{metal oxide} + \text{carbon} \rightarrow \text{metal} + \text{carbon dioxide}$$

For example:

$$\text{lead oxide} + \text{carbon} \rightarrow \text{lead} + \text{carbon dioxide}$$
$$2PbO + C \rightarrow 2Pb + CO_2$$

We call the removal of oxygen in this way a **reduction reaction**.

e) What do chemists mean by a reduction reaction?

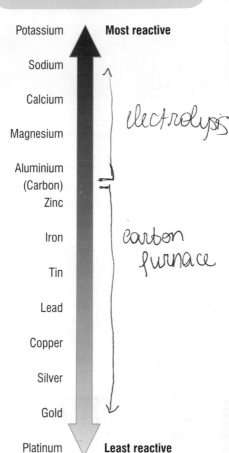

Potassium — **Most reactive**
Sodium
Calcium
Magnesium
Aluminium
(Carbon)
Zinc
Iron
Tin
Lead
Copper
Silver
Gold
Platinum — **Least reactive**

electrolysis

carbon furnace

Figure 4 This reactivity series shows how reactive each element is compared to the other elements

PRACTICAL

Reduction by carbon

Heat some copper oxide with carbon powder strongly in a test tube.

Empty the contents into an evaporating dish.

You can repeat the experiment with lead oxide and carbon if you have a fume cupboard to work in.

● Explain your observations. Include a word equation or balanced symbol equation.

SUMMARY QUESTIONS

1 Copy and complete using the words below:

   crust   extracted   native   reduction

   Metals come from the Earth's .... Some metals are very unreactive and are found in their ...... state. Metals, such as zinc and iron, are found combined with oxygen and can be ...... using chemical reactions. These are known as ...... reactions, as oxygen is removed from the oxide.

2 Make a list of all the metal objects found in your classroom or at home. Try to name the metal(s) used to make each object.

3 Platinum is never found combined with oxygen. What does this tell you about its reactivity? Give a use of platinum that depends on this property.

4 Zinc oxide (ZnO) can be reduced to zinc by heating it in a furnace with carbon. Carbon monoxide (CO) is given off in the reaction. Write a word equation and a balanced equation for the reduction of zinc oxide.

KEY POINTS

1 The Earth's crust contains many different elements.
2 A metal ore contains enough of the metal to make it economically worth extracting the metal.
3 We can find gold and other unreactive metals in their native state.
4 The reactivity series is useful in deciding the best way to extract a metal from its ore.
5 Metals more reactive than carbon cannot be extracted from their ores using carbon.

C1a 2.2 Extracting iron

LEARNING OBJECTIVES

1 What are the raw materials for making iron?
2 How is iron ore reduced?

Figure 1 Making iron is a hot, dirty process – and it can be quite spectacular too!

Extracting iron from its ore is a huge industry. Iron is the second most common metal in the Earth's crust. Iron ore contains iron combined with oxygen. Iron is less reactive than carbon. So we can extract iron by using carbon to remove oxygen from the iron oxide in the ore.

We extract iron using a **blast furnace**. This is a large container made of steel. It is lined with fireproof bricks to withstand the high temperatures inside. There are solid raw materials which we use in the blast furnace, as well as lots of air. They are:

● **Haematite** – this is the most common iron ore. It contains mainly iron(III) oxide (Fe_2O_3) and sand.
● **Coke** – this is made from coal and is almost pure carbon. It will provide the reducing agent to remove the oxygen from iron(III) oxide.

We also add limestone to remove impurities.

a) List the three solid raw materials for making iron.

Hot air is blown into the blast furnace. This makes the coke burn, which heats the furnace and forms carbon dioxide gas.

$$C + O_2 \rightarrow CO_2$$

At the high temperatures in the blast furnace, this carbon dioxide reacts with more coke to form carbon monoxide gas.

$$CO_2 + C \rightarrow 2CO$$

The carbon monoxide reacts with the iron oxide, removing its oxygen, and reducing it to molten iron. This flows to the bottom of the blast furnace.

$$Fe_2O_3 + 3CO \rightarrow 2Fe + 3CO_2$$

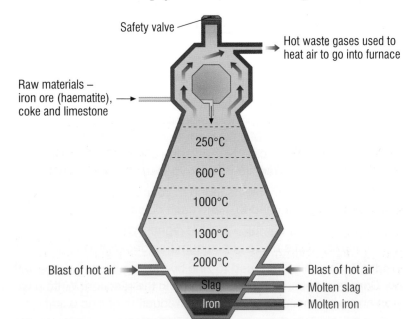

Safety valve

Hot waste gases used to heat air to go into furnace

Raw materials – iron ore (haematite), coke and limestone

250°C
600°C
1000°C
1300°C
2000°C

Blast of hot air

Blast of hot air

Slag

Molten slag

Iron

Molten iron

Figure 2 A blast furnace produces molten iron. Impurities from the iron ore are removed as slag.

Some of the molten iron is left to solidify in moulds – we call this **cast iron**. It contains about 96% iron. Most of the iron is kept molten to be turned into **steel**.

We make steel by removing more of the impurities from the iron. Then we mix it with other elements to change its properties.

b) What is the name of the substance that reduces the iron oxide to iron?

Iron – an important metal

Iron played a vital part in the Industrial Revolution, which happened in Britain during the 1700s and 1800s. Three generations of the same family – all called Abraham Darby – improved and developed new ways of making iron.

Up until this time charcoal had been used as the source of carbon for reducing the iron oxide. The Darby family replaced this with coke. They developed the blast furnace as a way of making iron continuously rather than in batches.

Most of the world's steel is now made in Asia. That's because the cost of making it is much lower than in Europe and North America. Iron and steel-making used to employ many thousands of people in the UK but now employs far fewer. Other jobs for these people have had to be made in other industries.

Figure 3 The youngest Darby built the world's first cast-iron bridge in 1779. It still spans the River Severn at Coalbrookdale in Shropshire. Coalbrookdale is now much better known as Ironbridge Gorge.

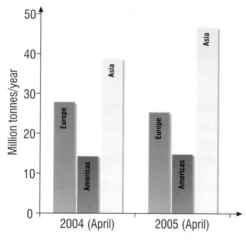

Figure 4 World steel output in April 2004 and April 2005

SUMMARY QUESTIONS

1 Copy and complete using the words below:

carbon   coke   haematite   limestone

We can extract iron from iron ore using .... The most common type of iron ore is .... Other raw materials for making iron are ...... and ....

2 How is steel different from cast iron?

3 a) Some of the iron(III) oxide (Fe_2O_3) in the blast furnace is reduced by carbon, giving off carbon dioxide. Write a word equation for this reaction.

b) Write a balanced symbol equation for the reaction described in part a).

KEY POINTS

1 We extract iron from iron ore by reducing it with carbon in a blast furnace.
2 The solid raw materials used to make iron are iron ore (haematite), coke and limestone.
3 Molten iron is tapped off from the bottom of the blast furnace.

C1a 2.3   Properties of iron and steels

The iron produced by a blast furnace, called pig iron, is not very useful. It is about 96% iron and contains impurities, mainly of carbon. This makes pig iron very brittle. However, we can treat the iron from the blast furnace to remove some of the carbon.

Figure 1 The iron which has just come out of a blast furnace contains about 96% iron. The main impurity is carbon.

If we remove all of the carbon and other impurities from pig iron, we get pure iron. This is very soft and easily-shaped, but it is too soft for most uses. If we want to make iron really useful we have to make sure that it contains tiny amounts of other elements, including carbon and certain metals.

We call a metal that contains other elements an **alloy**.

Iron that has been alloyed with other elements is called **steel**. By adding elements in carefully controlled amounts, we can change the properties of the steel.

a) Why is iron from a blast furnace very brittle?
b) Why is pure iron not very useful?
c) How do we control the properties of steel?

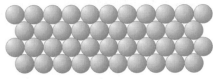

Iron

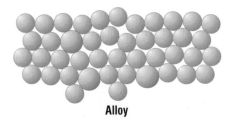

Alloy

Figure 2 The atoms in pure iron are arranged in layers which can easily slide over one another. In alloys the layers cannot slide so easily because atoms of other elements change the regular structure.

Look at Figure 2:
The atoms in pure iron are arranged in layers. Because of this regular arrangement, the atoms can slide over one another very easily. This is why pure iron is so soft and easily shaped.

Steels

Steel is not a single substance. There are lots of different types of steel. All of them are alloys of iron with carbon and/or other elements.

The simplest steels are the *carbon steels*. We make these by alloying iron with small amounts of carbon (from 0.03% to 1.5%). These are the cheapest steels to make. We use them in many products, like the bodies of cars, knives, machinery, ships, containers and structural steel for buildings.

Often these carbon steels have small amounts of other elements in them as well. High carbon steel, with a relatively high carbon content, is very strong but brittle. On the other hand, low carbon steel is soft and easily shaped. It is not as strong, but it is much less likely to shatter.

Mild steel is one type of low carbon steel. It contains less than 0.1% carbon. It is very easily pressed into shape. This makes it particularly useful in mass production, for example, making car bodies.

Low-alloy steels are more expensive than carbon steels because they contain between 1% and 5% of other metals. Examples of the metals used include nickel, chromium, manganese, vanadium, titanium and tungsten. Each of these metals gives a steel that is well-suited for a particular use. We use nickel–steel alloys to build long-span bridges, bicycle chains and military armour-plating. That's because they are very resistant to stretching forces. Tungsten steel operates well under very hot conditions so it is used to make high-speed tools.

Even more expensive are the *high-alloy steels*. These contain a much higher percentage of other metals. For example, chromium steels have between 12% and 15% chromium mixed with the iron, and often some nickel is mixed in too – this provides strength and chemical stability.

These chromium–nickel steels are more commonly known as *stainless steels*. We use them to make cooking utensils and cutlery. They are also used to make chemical reaction vessels because they combine hardness and strength with great resistance to corrosion. Unlike most other steels, they do not rust!

Figure 3 The properties of steel alloys make them ideal for use in suspension bridges

GET IT RIGHT!

Know how the hardness of steels is related to their carbon content.

SUMMARY QUESTIONS

1 Complete the following sentences using the terms below:

      carbon steel   pure iron   steel

If all of the carbon and other impurities are removed from pig iron we get
....
Iron that has been alloyed with other elements is called ....
Iron that has been alloyed with a little carbon is called ....

2 a) What is the difference between high-alloy and low-alloy steels?
   b) Why are surgical instruments made from steel containing chromium and nickel?
   c) Make a table to summarise the composition and properties of different types of steel.
   d) Using diagrams, explain how alloying a metal with atoms of another element changes its properties.

KEY POINTS

1 Pure iron is too soft for it to be very useful.
2 Carefully controlled quantities of elements are added to iron to make alloys of steel with different properties.

43

C1a 2.4   Alloys in everyday use

As we saw on page 42, pure iron is not very useful because it is too soft. In order to use the iron, we must turn it into an alloy by adding other elements. Other metals must also be turned into alloys to make them as useful as possible.

Copper

Copper is a soft, reddish coloured metal. It conducts heat and electricity very well. We have used copper for thousands of years. In fact some people think that it was the first metal to be used by humans. Copper articles have been found which are well over ten thousand years old.

Just like pure iron, pure copper is very soft.

Figure 1 The Statue of Liberty in New York contains over 80 tonnes of copper

Bronze was probably the first alloy made by humans, about 5500 years ago. It is usually made by mixing copper with tin, but small amounts of other elements can be added as well. For example, we can add phosphorus. This gives the alloy properties which make it ideal to use for bearings where we want very low friction.

We make *brass* by alloying copper with zinc. Brass is much harder than copper and it is workable.

It can be hammered into sheets and bent into different shapes. This property is used to make musical instruments.

a) What are the names of two common alloys of copper?
b) Why are copper alloys more useful than pure copper?

Gold and aluminium

Just like copper and iron, we can make gold and aluminium harder by adding other elements. We usually alloy gold with copper and silver when we want to use it in jewellery. By varying the proportions of the two metals added we can get differently coloured 'gold' objects. They can vary from yellow to red, and even a shade of green!

Aluminium is a metal with a low density. It can be alloyed with a wide range of other elements – there are over 300 alloys of aluminium! These alloys have very different properties. We can use some to build aircraft while others can be used as amour plating on tanks and other military vehicles.

c) Apart from making gold harder, what else can alloying change?

d) What property of aluminium makes it useful for making alloys in the aircraft industry?

Smart alloys

Some alloys have a very special property. If we bend (or **deform**) them into a different shape and then heat them, they return to their original shape all by themselves. These alloys are sometimes called **smart alloys**. Their technical name is **shape memory alloys** (SMAs), which describes the way they behave. They seem to 'remember' their original shape!

We can use the clever properties of shape memory alloys in many ways. Some of the most interesting uses of SMAs have been in medicine. Doctors treating a badly broken bone can use smart alloys to hold the bones in place while they heal. They cool the alloy before it is wrapped around the broken bone. When it heats up again the alloy goes back to its original shape. This pulls the bones together and holds them while they heal. Dentists have made braces to push teeth into the right position using this technique.

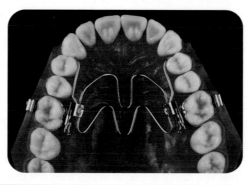

Figure 2 This dental brace pulls the teeth into the right position as it warms up. That's smart!

KEY POINTS

1 Copper, gold and aluminium are all alloyed with other metals to make them more useful.

2 We can control the properties of alloys by adding different amounts of different elements.

3 Smart alloys are also called shape memory alloys. When deformed they return to their original shape on heating.

4 Shape memory alloys can be used in medicine and dentistry.

SUMMARY QUESTIONS

1 Copy and complete using the words below:

aluminium   brass   bronze   smart   soft   thousands

Copper has been used by people for …… of years. Like pure iron, pure copper is too …… to be very useful. Copper can be alloyed with tin to make ……, and with zinc to make …… . There are over 300 alloys of …… . Some alloys can 'remember' their shape when they are heated after they have been bent – they are called …… alloys.

2 Why can aluminium alloys be used in so many different ways?

3 a) Explain the advantages of a dental brace made of 'smart' alloy over one made of a conventional alloy.

b) Do some research to find some other uses of smart alloys, explaining why they are used.

C1a 2.5   Transition metals

LEARNING OBJECTIVES

1 What are transition metals and why are they useful?
2 Why do we use so much copper?
3 How can we produce enough copper?

In the centre of the periodic table there is a large block of metallic elements. Here we find the elements called the **transition metals** or *transition elements*. Many of them have similar properties.

Like all metals, the transition metals are very good conductors of electricity and heat. They are also hard, tough and strong. Yet we can easily bend or hammer them into useful shapes. We say they are **malleable**. With the exception of mercury, which is a liquid at room temperature, the transition metals have very high melting points.

a) In which part of the periodic table do we find the transition metals?
b) Name *three* properties of these elements.

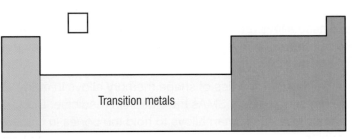

Figure 1 The transition metals

The properties of the transition metals mean that we can use them in many different ways. You will find them in buildings and in cars, trains and other types of transport. Their strength makes them useful as *construction materials*. We use them in heating systems and for electrical wiring because heat and electricity pass through them easily.

One of the transition metals is copper. Although copper is not particularly strong, we can bend and shape it easily. It also conducts electricity and heat very well and it does not react with water. So it is ideal where we need pipes that will carry water or wires that will conduct electricity.

c) Why do we use transition metals so much?
d) What makes copper so useful?

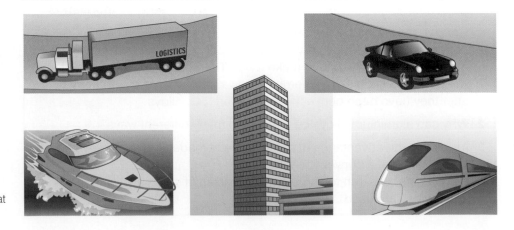

Figure 2 Transition metals are used in many different ways because of their properties. Copper is particularly useful because it is such a good conductor of heat and electricity.

Figure 3 Mining copper ores can leave huge scars on the landscape. This is called open cast mining.

We extract copper from **copper ore**. There are two main methods used to remove the copper from the ore. In one method we use sulfuric acid to produce copper sulfate solution, before extracting the copper.

The other process is called **smelting**. We heat copper ore very strongly in air to produce crude copper. Then we use the impure copper as anodes in electrolysis cells to make pure copper. 85% of copper is still produced by smelting.

e) What chemical do we use to treat copper ore in order to form copper sulfate?

Processing copper ore uses huge amounts of electricity, and costs a lot of money. If we have to smelt an ore, the heating also requires a lot of energy.

New ways to extract copper

Instead of extracting copper using chemicals, heat and electricity, scientists are developing new ways to do the job. They can now use bacteria, fungi and even plants to help extract copper.

If we could extract metals like this on a large scale, it could be a lot cheaper than the way we do it now. It could be a lot 'greener' too.

We would also be able to extract copper from ores which contain very little copper. At the moment it is too expensive to process these 'low grade' ores using conventional methods.

NEXT TIME YOU...

... take a walk in your garden or go onto the school field, think about the transition metals under your feet. Although they might be too expensive to extract now, new ways of extracting metals could mean that you are literally standing on a goldmine!

Figure 4 In Australia Dr Jason Plumb looks for bacteria that can extract metals from ores. His search takes him to some exciting places – including volcanoes!

SUMMARY QUESTIONS

1 Write a few words describing the following: a) transition metals, b) properties of copper, c) smelting.

2 Silver and gold are transition metals that conduct electricity even better than copper. Why do we use copper to make electric cables instead of either of these metals?

3 a) Explain briefly two ways of extracting copper metal.
   b) Explain the advantages of extracting copper using bacteria or fungi rather than the way most is extracted now.

KEY POINTS

1 The transition metals are found in the middle block of elements in the periodic table.
2 Transition metals have properties that make them useful for building and making things.
3 Copper is a very useful transition metal because of its high conductivity.
4 Scientists are looking for new ways to extract copper that will use less energy.

C1a 2.6   Aluminium and titanium

Although they are very strong, many metals are also very dense. This means that we cannot use them if we want to make something that has to be both strong and light. Examples are alloys for making an aeroplane or the frame of a racing bicycle.

Where we need metals which are both strong and light, *aluminium* and *titanium* fit the bill. They are also metals which do not *corrode*.

Aluminium is a silvery, shiny metal which is surprisingly light – it has a relatively low density for a metal. It is an excellent conductor of heat and electricity. We can also shape it and draw it into wires very easily.

Although aluminium is a relatively reactive metal, it does not corrode easily. This is because the aluminium atoms at its surface immediately react with the oxygen in air to form a thin layer of tough aluminium oxide. This layer stops any further corrosion taking place.

Aluminium is not a particularly strong metal, but we can use it to form alloys. These alloys are harder, more rigid and stronger than pure aluminium.

Figure 1 We use aluminium alloys to make aircraft because of their combination of lightness and strength

As a result of these properties, we use aluminium to make a whole range of goods. These range from cans, cooking foil and saucepans through to high-voltage electricity cables, aeroplanes and space vehicles.

a) Why does aluminium resist corrosion?
b) How do we make aluminium stronger?

Titanium is a silvery-white metal, which is very strong and very resistant to corrosion. Like aluminium it has an oxide layer that protects it. Although it is denser than aluminium, it is less dense than most transition metals.

Titanium has a very high melting point – about 1660°C – so we can use it at very high temperatures.

We use it instead of steel and aluminium in the bodies of high-performance aircraft and racing bikes. Here its combination of relative low density and strength is important. We also use titanium to make parts of jet engines because it keeps its strength at high temperatures.

The strength of titanium at high temperatures makes it very useful in nuclear reactors. In reactors we use it to make the pipes and other parts that must stand up to high temperatures. Titanium performs well under these conditions, and its strong oxide layer means that it resists corrosion.

Another use of titanium is also based on its strength and resistance to corrosion – replacement hip joints.

c) Why does titanium resist corrosion?
d) What properties make titanium ideal to use in jet engines and nuclear reactors?

We extract aluminium using electrolysis. Because it is a reactive metal we cannot use carbon to displace it from its ore. Instead we extract aluminium by passing an electric current through molten aluminium oxide at high temperatures.

Titanium is not particularly reactive, so we could produce it by displacing it from its ore with carbon. But carbon reacts with the titanium making it very brittle. So we have to use a more reactive metal. We use sodium or magnesium to do this. However, we have to produce both sodium and magnesium by electrolysis in the first place.

The problem with using electrolysis to extract these metals is that it is very expensive. That's because we need to use high temperatures and a great deal of electricity.

In the UK each person uses around 8 kg of aluminium every year. This is why it is important to **recycle** aluminium. It saves energy, since recycling aluminium does not involve electrolysis.

e) Why do we need electricity to make aluminium and titanium?
f) Why does recycling aluminium save electricity?

Figure 2 We can use titanium inside the body as well as outside. Because it has a low density, is strong and does not corrode, we can use titanium to make alloys that are excellent for artificial joints. These are artificial hip joints, used to replace a natural joint damaged by disease or wear and tear.

SUMMARY QUESTIONS

1 Copy and complete using the words below:

   corrode   electrolysis   expensive   light   oxide   reactive   strong

   Aluminium and titanium alloys are useful as they are ...... and ....
   Although aluminium is reactive, it does not ...... because its surface is
   coated with a thin layer of aluminium .... Titanium does not corrode
   because it is not very ...... and also has its oxide layer to protect it.
   ...... is used in the extraction of both metals from their ores which makes
   them ....

2 Why is titanium used to make artificial hip joints?

3 Each person in the UK uses about 8 kg of aluminium each year.

   a) Recycling 1 kg of aluminium saves about enough energy to run a
      small electric fire for 14 hours. If you recycle 50% of the aluminium
      you use in one year, how long could you run a small electric fire on
      the energy you have saved?
   b) Explain the benefits of recycling aluminium.

KEY POINTS

1 Aluminium and titanium are useful because they resist corrosion.
2 Aluminium and titanium are expensive because extracting them from their ores involves many stages in the processes and requires large amounts of energy.
3 Recycling aluminium is important because we need to use much less energy to produce 1 kg of recycled aluminium than we use to extract 1 kg of aluminium from its ore.

C1a 2.7   Using metals

Metals and society

The way we use metals has literally changed the world. Since ancient times metals have enabled us to do things we could never have done without them. These range from making tools to generating electricity; from creating jewellery to flying in aeroplanes.

The timeline below shows some of the main points in the history of the use of metals, up to the beginning of the 20th century.

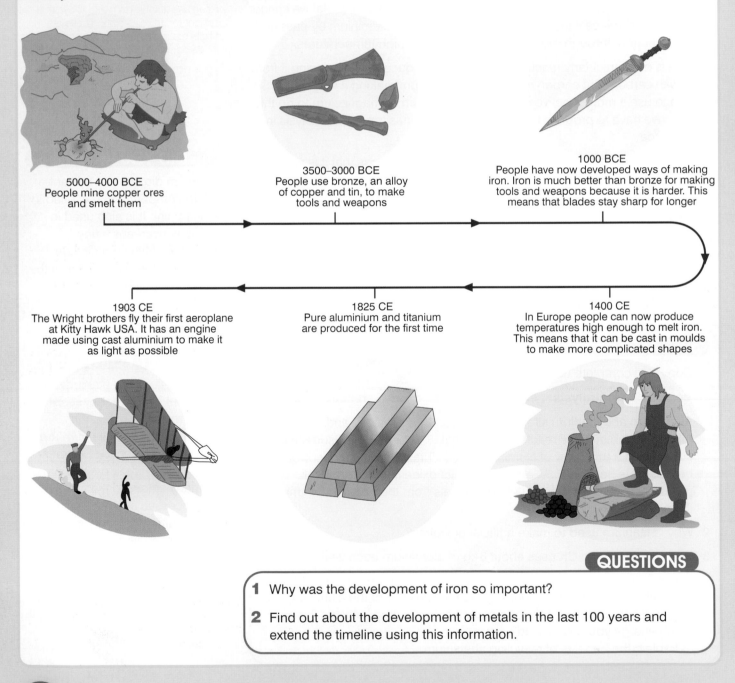

5000–4000 BCE
People mine copper ores and smelt them

3500–3000 BCE
People use bronze, an alloy of copper and tin, to make tools and weapons

1000 BCE
People have now developed ways of making iron. Iron is much better than bronze for making tools and weapons because it is harder. This means that blades stay sharp for longer

1903 CE
The Wright brothers fly their first aeroplane at Kitty Hawk USA. It has an engine made using cast aluminium to make it as light as possible

1825 CE
Pure aluminium and titanium are produced for the first time

1400 CE
In Europe people can now produce temperatures high enough to melt iron. This means that it can be cast in moulds to make more complicated shapes

QUESTIONS

1 Why was the development of iron so important?

2 Find out about the development of metals in the last 100 years and extend the timeline using this information.

Mining with plants

As they grow, plants absorb dissolved chemicals from the ground in the water that they take into their roots. Some plants absorb a lot of one certain chemical, and scientists have found that we can use these plants to help us extract metals from soil.

The technology that we need to do this is quite simple. So we can use plants to mine metals where it would be far too expensive to do it in any other way. Extracting metals like this is called *phytomining*.

QUESTIONS

3 Why can we use phytomining when it would be too expensive to extract metals in other ways?

4 There is a long time delay between planting the crop and extracting the metal. What might happen if the price of the metal falls before the crop can be harvested?

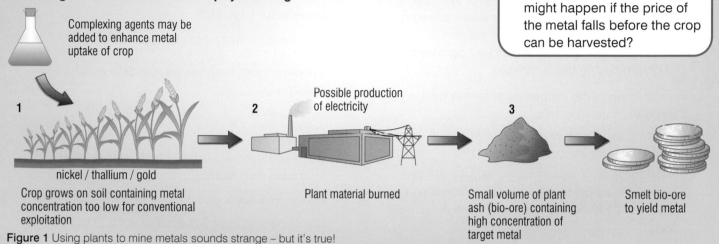

Complexing agents may be added to enhance metal uptake of crop

1 nickel / thallium / gold

Crop grows on soil containing metal concentration too low for conventional exploitation

Possible production of electricity

2 Plant material burned

3 Small volume of plant ash (bio-ore) containing high concentration of target metal

Smelt bio-ore to yield metal

Figure 1 Using plants to mine metals sounds strange – but it's true!

Recycling fridges

It is nearly always cheaper for us to recycle metals than to extract new metals from their ores. This is especially true for metals like aluminium, where we need to use large amounts of energy to extract them. But sometimes it is not easy to recycle metals because they are combined with other materials. As an example, look at the problems of recycling fridges.

ACTIVITY

Many people do not realise how important it is to recycle old fridges and other household equipment.
Design a poster to be used in a campaign to persuade people to recycle their old fridge, washing machine, and so on, rather than simply dumping them.

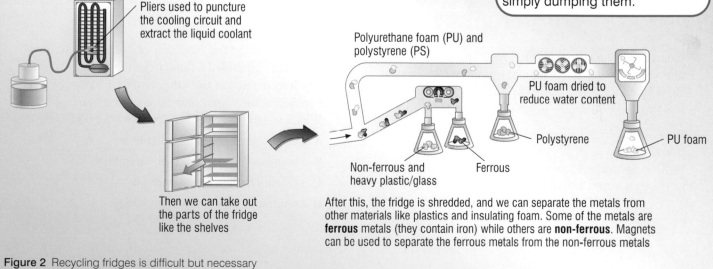

First, we need to remove the chemicals in the cooling system of the fridge

Pliers used to puncture the cooling circuit and extract the liquid coolant

Polyurethane foam (PU) and polystyrene (PS)

PU foam dried to reduce water content

Polystyrene

PU foam

Non-ferrous and heavy plastic/glass

Ferrous

Then we can take out the parts of the fridge like the shelves

After this, the fridge is shredded, and we can separate the metals from other materials like plastics and insulating foam. Some of the metals are **ferrous** metals (they contain iron) while others are **non-ferrous**. Magnets can be used to separate the ferrous metals from the non-ferrous metals

Figure 2 Recycling fridges is difficult but necessary

SUMMARY QUESTIONS

1 Write simple definitions for the following terms:

   a) metal ore

   b) native state

   c) reduction reaction.

2 Gold is a very dense, unreactive metal. Old-fashioned gold prospectors used to 'pan' for gold in streams by scooping up small stones from the stream bed and washing them in water, allowing them slowly to get washed out of the pan. Using words and diagrams, explain how they could find gold by using this technique.

3 We can change the properties of metals by alloying them with other elements.

   a) Write down **three** ways that a metal alloy may be different from the pure metal.

   b) Choose **one** of these properties. Use the two diagrams below to help you to explain why a metal alloy behaves differently to the pure metal.

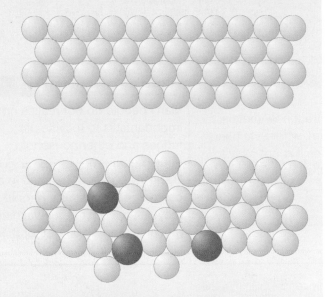

4 One of your fellow students says: 'There is more aluminium in the Earth's crust than any other metal. So why should we bother recycling it? How would you argue against this point of view?

5 One use of smart alloys is to make spectacle frames.
Write down **one** advantage and **one** disadvantage of using a smart alloy like this.

EXAM-STYLE QUESTIONS

1 This question is about the uses of these metals:

   A aluminium   **B** copper

   C gold   **D** iron

   Which of these metals is used

   (a) as the main metal in alloys to build aircraft?

   (b) in alloys to make jewellery?

   (c) to make all steels?

   (d) to make water pipes and electrical wiring?   (4)

2 Choose a metal from the list **A** to **D** to match each description.

   A aluminium   **B** chromium

   C gold   **D** titanium

   (a) A metal that is strong at high temperatures and resists corrosion.

   (b) An unreactive metal found native in the Earth.

   (c) This metal has a low density and is extracted by electrolysis.

   (d) This metal is mixed with iron to make high alloy steels.   (4)

3 Use words from the list **A** to **D** to complete the word equations.

   A copper   **B** iron

   C sodium   **D** water

   (a) copper oxide + sulfuric acid → copper sulfate + ...

   (b) copper sulfate + iron → ...... + iron sulfate

   (c) iron oxide + carbon → ...... + carbon dioxide

   (d) titanium + ... → titanium + sodium
       chloride   chloride   (4)

4 A student tested the flexibility of four different alloy rods. She suspended a mass from the end of the rods which were fixed at the other end to the edge of a bench. She measured how far each rod bent.
Which words describe the 'distance the rod bent'?

   A a categoric, independent variable.

   B a continuous, independent variable.

   C a categoric, dependent variable.

   D a continuous, dependent variable.   (1)

5 Name the types of substance described in each part of this question.

(a) These elements are hard, tough and strong, conduct heat and electricity well and are found in the middle of the periodic table.

(b) These rocks contain enough metal to make it worth extracting.

(c) This is a metal that contains other elements to give it specific properties.

(d) These materials are smart because they can return to their original shape when heated and are used by surgeons to hold broken bones while healing. (4)

6 Iron is extracted from iron oxide by removing oxygen.

(a) What name is given to a reaction in which oxygen is removed from a compound? (1)

(b) Name an element that could be used to remove oxygen from iron oxide. (1)

(c) Write a word equation for the reaction that would take place in (b). (2)

7 Titanium is used to make replacement hip joints. One reason why titanium can be used in this way is that it resists corrosion.

(a) How is titanium protected from corrosion? (1)

(b) Suggest **two** other properties of titanium that make it suitable for this use. (2)

8 Most of the world's steel is now made in Asia.

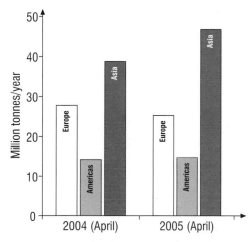

Suggest **two** reasons why it costs less to make steel in Asia than in Europe. (2)

9 New methods using bacteria, fungi and plants are being developed to extract copper. Suggest **three** reasons why these new methods have been developed. (3)

HOW SCIENCE WORKS QUESTIONS

How hard is gold?

The following was overheard in a jeweller's shop:

'I would like to buy a 24 carat gold ring for my husband.'

'Well madam, we would advise that you buy one which is lower carat gold. It looks much the same but the more gold there is, the less hard it is.'

Is this actually the case? Let's have a look scientifically at the data.

Pure gold is 24 carat. A carat is a twenty-fourth, so $24 \times 1/24 = 1$ or pure gold. So a 9 carat gold ring will have 9/24ths gold and 15/24ths of another metal, probably copper or silver. Most 'gold' sold in shops is therefore an alloy. How hard the 'gold' is will depend on the amount of gold and on the type of metal used to make the alloy.

Here is some data on the alloys and the maximum hardness of 'gold'.

Gold alloy (carat)	Maximum hardness (BHN)
9	170
14	180
18	230
22	90
24	70

a) The shop assistant said that 'the more gold there is, the less hard it is.' Was this based on science or was it hearsay? Explain your answer. (2)

b) In this investigation which is the independent variable? (1)

c) Which of the following best describes the hardness of the alloy?
   i) continuous
   ii) discrete
   iii) categoric
   iv) ordered. (1)

d) Plot a graph of the results. (3)

e) What is the pattern in your results? (2)

f) You might have expected that the 9 carat gold was much harder than the 14 or the 18 carat gold, but it isn't. Can you offer an explanation for this? (1)
(Clue – is there an uncontrolled variable lurking around here?)

C1a 3.1   Fuels from crude oil

Some of the 21st century's most important chemistry involves chemicals that are made from crude oil. These chemicals play a major part in our lives. We use them as fuels in our cars, to warm our homes and to make electricity.

Fuels are important because they keep us warm and on the move. So when oil prices rise, it affects us all. Countries that produce crude oil can have an affect on the whole world economy by the price they charge for their oil.

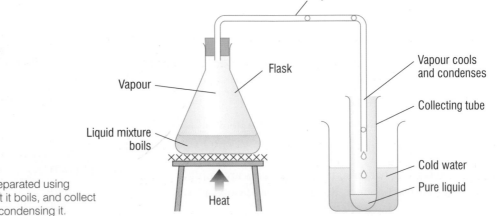

Figure 1 The price of nearly everything we buy is affected by oil because the cost of moving goods to the shops affects the price we pay for them

a) Why is oil so important?

Crude oil

Crude oil is a dark, smelly liquid, which is a *mixture* of lots of different chemical compounds. A mixture contains two or more elements or compounds that are not chemically combined together.

Crude oil straight out of the ground is not much use. There are too many substances in it, all with different boiling points. Before we can use crude oil, we have to separate it into its different substances. Because the properties of substances do not change when they are mixed, we can separate mixtures of substances in crude oil by using *distillation*. Distillation separates liquids with different boiling points.

Figure 2 Mixtures of liquids can be separated using distillation. We heat the mixture so that it boils, and collect the vapour that forms by cooling and condensing it.

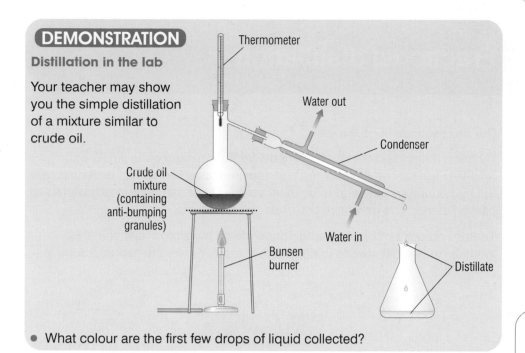

DEMONSTRATION

Distillation in the lab

Your teacher may show you the simple distillation of a mixture similar to crude oil.

Thermometer

Water out

Condenser

Crude oil mixture (containing anti-bumping granules)

Water in

Bunsen burner

Distillate

● What colour are the first few drops of liquid collected?

b) What is crude oil?

c) Why can we separate crude oil using distillation?

Nearly all of the compounds in crude oil are made from atoms of just two chemical elements – hydrogen and carbon. We call these compounds **hydrocarbons**. Most of the hydrocarbons in crude oil are **alkanes**. You can see some examples of alkane molecules in Figure 3.

Another way of writing these alkane molecules is like this:

CH_4 (methane); C_2H_6 (ethane); C_3H_8 (propane); C_4H_{10} (butane); C_5H_{12} (pentane).

Can you see a pattern in the formulae of the alkanes? We can write the general formula for alkane molecules like this:

$$C_nH_{2n+2}$$

which means that 'for every n carbon atoms there are $(2n + 2)$ hydrogen atoms'. For example, if an alkane contains 25 carbon atoms its formula will be $C_{25}H_{52}$.

We describe alkanes as **saturated** hydrocarbons. This means that they contain as many hydrogen atoms as possible in each molecule. We cannot add any more.

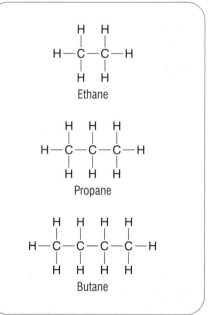

Figure 3 We can represent alkanes like this, showing all of the atoms in the molecule. The line between two atoms in the molecule is the chemical bond holding them together.

SUMMARY QUESTIONS

1 Copy and complete using the words below:

    carbon   distillation   hydrocarbons   hydrogen   mixture

    Crude oil is a …… of compounds. Many of these contain only atoms of …… and …… . They are called …… . The compounds in crude oil can be separated using …… .

2 Why is crude oil not very useful before we have processed it?

3 a) Write down the formula of the alkanes which have 6, 7 and 8 carbon atoms. Then find out their names.

    b) How many carbon atoms are there in an alkane which has 30 hydrogen atoms?

KEY POINTS

1 Crude oil is a mixture of many different compounds.

2 Many of the compounds in crude oil are hydrocarbons – they contain only carbon and hydrogen.

3 Alkanes are saturated hydrocarbons. They contain as much hydrogen as possible in their molecules.

C1a 3.2   Fractional distillation

The compounds in crude oil

Hydrocarbon molecules can be very different. Some are quite small, with relatively few carbon atoms and short chains. These short-chain molecules are the hydrocarbons that tend to be most useful. Other hydrocarbons have lots of carbon atoms, and may have branches or side chains.

The boiling point of a hydrocarbon depends on the size of its molecules. We can use the differences in boiling points to separate the hydrocarbons in crude oil.

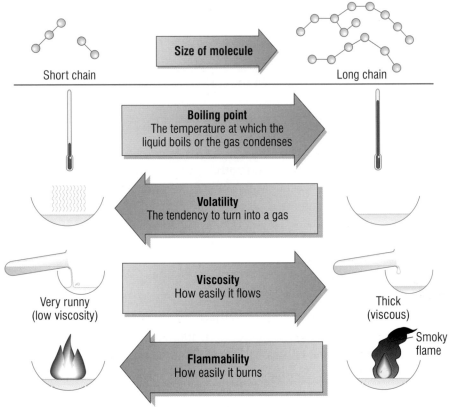

Figure 1 The properties of hydrocarbons depend on the chain length of their molecules.

a) How does the length of the hydrocarbon chain affect:
  (i) the boiling point, and
  (ii) the viscosity (thickness) of a hydrocarbon?
b) A hydrocarbon catches fire very easily. Is it likely to have molecules with long hydrocarbon chains or short ones?

Refining crude oil

We separate out crude oil into hydrocarbons with similar properties, called *fractions*. We call this process **fractional distillation**. Each hydrocarbon fraction contains molecules with similar numbers of carbon atoms. Each of these fractions boils at different temperatures. That's because of the different numbers of carbon atoms in their molecules.

Crude oil is fed in near the bottom of a tall tower (a fractionating column) as hot vapour. The tower is kept very hot at the bottom and much cooler at the top, so the temperature decreases going up the column. The gases in the column condense when they reach their boiling points, and the different fractions are collected at different levels.

Hydrocarbons with the smallest molecules have the lowest boiling points, so they are collected at the cool top of the tower. At the bottom of the tower the fractions have high boiling points. They cool to form very thick liquids or solids at room temperature.

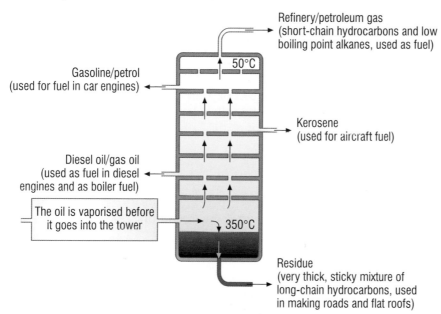

Refinery/petroleum gas
(short-chain hydrocarbons and low boiling point alkanes, used as fuel)

50°C

Gasoline/petrol
(used for fuel in car engines)

Kerosene
(used for aircraft fuel)

Diesel oil/gas oil
(used as fuel in diesel engines and as boiler fuel)

The oil is vaporised before it goes into the tower

350°C

Residue
(very thick, sticky mixture of long-chain hydrocarbons, used in making roads and flat roofs)

Figure 2 We use fractional distillation to turn the mixture of hydrocarbons in crude oil into fractions, each containing compounds with similar boiling points

Once we have collected them, the fractions need more refining before they can be used.

There are many different types of crude oil. For example, crude oil from Venezuela contains many long-chain hydrocarbons. It is very dark and thick and we call it 'heavy' crude. Other countries, such as Nigeria and Saudi Arabia, produce crude oil which is much paler in colour and more runny. This is 'light' crude.

Light crude oil contains many more of the smaller molecules we can use as fuels and for making other chemicals. So light crude oils cost more to buy than the heavier crude oils which contain a higher percentage of the larger hydrocarbons.

c) Why is 'light' crude oil likely to cost more than 'heavy' crude?

SUMMARY QUESTIONS

1 Copy and complete using the words below:

    easily   fractional distillation   fractions   high
              mixture   viscosity

Crude oil is a ...... of many different hydrocarbons. We can separate crude oil into different ...... using .... Hydrocarbon molecules with many carbon atoms tend to have ...... boiling points and .... Hydrocarbon molecules with few carbon atoms catch fire ....

2 a) Explain the steps involved in the fractional distillation of crude oil.
  b) Make a table to summarise how the properties of hydrocarbons depend on the size of their molecules.

KEY POINTS

1 We separate crude oil into fractions using fractional distillation.
2 Properties of each fraction depend on the size of the hydrocarbon molecules.
3 Lighter fractions make better fuels.

C1a 3.3   Burning fuels

LEARNING OBJECTIVES

1 What do we produce when we burn fuels?
2 When we burn fuels, how do changes in conditions affect what is produced?
3 What pollutants are produced when we burn fuels?

Figure 1 On a cold day we can often see the water produced when fossil fuels burn

As we saw on page 57, the lighter fractions produced from crude oil are very useful as fuels. When hydrocarbons burn in plenty of air they produce two new substances – carbon dioxide and water.

For example, when propane burns we can write:

$$propane + oxygen \rightarrow carbon\ dioxide + water$$

or

$$C_3H_8 + 5O_2 \rightarrow 3CO_2 + 4H_2O$$

Notice how we need five molecules of oxygen for the propane to burn. This makes three molecules of carbon dioxide and four molecules of water. The equation is **balanced**!

PRACTICAL

Products of combustion

We can test the products given off when hydrocarbon burns as shown in Figure 2.

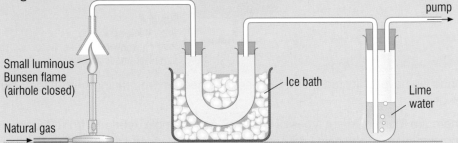

Small luminous Bunsen flame (airhole closed)

Natural gas

To water pump

Ice bath

Lime water

Figure 2 Testing the products formed when hydrocarbons burn

● What happens to the lime water?   ● What collects in the U-tube?

a) What are the names of the two substances produced when hydrocarbons burn in plenty of air?
b) Write a balanced equation for methane (CH_4) burning in plenty of air.

All fossil fuels – oil, coal and natural gas – produce carbon dioxide and water when they burn in plenty of air. But as well as hydrocarbons, these fuels also contain other substances. These produce different compounds when we burn the fuel, and this can cause problems for us.

Impurities of sulfur cause us major problems. All fossil fuels contain at least some of this element, which reacts with oxygen when we burn the fuel. It forms a gas called **sulfur dioxide**. This gas is poisonous. It is also acidic. This is bad for the environment, as it is a cause of acid rain.

Sulfur dioxide can also cause engine corrosion.

c) When hydrocarbons burn, what element present in the impurities in a fossil fuel may produce sulfur dioxide?
d) Why is it bad if sulfur dioxide is produced?

When we burn hydrocarbons in a car engine, even more substances can be produced. When there is not enough oxygen inside the cylinders of the engine, we get *incomplete combustion*. Instead of all the carbon in the fuel turning into carbon dioxide, we also get *carbon monoxide* (CO).

Carbon monoxide is a poisonous gas. Your red blood cells pick up this gas and carry it around in your blood instead of oxygen. So even quite small amounts of carbon monoxide gas are very bad for you.

The high temperature inside an engine also allows the nitrogen and oxygen in the air to react together. This makes *nitrogen oxides*, which are poisonous and which can trigger some people's asthma. They also make acid rain.

Diesel engines burn hydrocarbons with much bigger molecules than petrol engines. When these big molecules react with oxygen in an engine they do not always burn completely. Tiny particles containing carbon and unburnt hydrocarbons are produced. These *particulates* get carried into the air of our towns and cities. We do not understand fully what particulates may do when we breathe them in. However, scientists think that they may damage the cells in our lungs and perhaps even cause cancer.

DID YOU KNOW...

. . . in 1996 a three-wheeled diesel car set a record for the lowest amount of fuel used to go a fixed distance? The car could travel 568 miles on one gallon (4.5 litres) of fuel!

Figure 3 The effect of many cars in a small area. Under the right weather conditions smog can be formed. This is a mixture of SMoke and fOG. Smog formed from car pollution contains many different types of chemicals which can be harmful to us.

Engine
If you have a car with a modern engine which you have serviced regularly, it will produce much less pollution than an old, badly-serviced car.

Fuel tank
When we fill up our cars some hydrocarbons escape into the atmosphere. Fuels contain hydrocarbons that are poisonous and which may cause cancer.

Exhaust
As well as carbon dioxide and water vapour, exhaust gases may contain carbon monoxide, nitrogen oxides, sulfur dioxide and tiny particles of unburnt carbon and hydrocarbons.

SUMMARY QUESTIONS

1 Copy and complete using the words below:

   carbon   carbon dioxide   nitrogen   particulates   sulfur   water

   When we burn hydrocarbons in plenty of air .... and ...... are made. As well as these compounds other substances like ...... dioxide may be made. Other pollutants that may be formed include ...... oxides, ... monoxide and ....

2 a) Natural gas is mainly methane (CH_4). Write a balanced equation for the complete combustion of methane.
   b) When natural gas burns in a faulty gas heater it can produce carbon monoxide (and water).
      Write a balanced equation to show this reaction.
   c) Explain how i) sulfur dioxide ii) nitrogen oxides and iii) particulates are produced when fuels burn in vehicles.

KEY POINTS

1 When we burn hydrocarbon fuels in plenty of air they produce carbon dioxide and water.
2 Impurities in fuels may produce other substances which may be poisonous and/or which may cause pollution.
3 Changing the conditions in which we burn fuels may change the products that are made.

C1a 3.4   Cleaner fuels

LEARNING OBJECTIVES

1 When we burn fuels, what are the consequences?
2 What can we do to reduce the problems?

When we burn hydrocarbons, as well as producing carbon dioxide and water, we produce other compounds. Many of these are not good for the environment, and can affect our health.

Pollution from our cars does not stay in one place but spreads through the whole of the Earth's atmosphere. For a long time the Earth's atmosphere seemed to cope with all this pollution. But the increase in the number of cars in the last 50 years means that pollution is a real concern now.

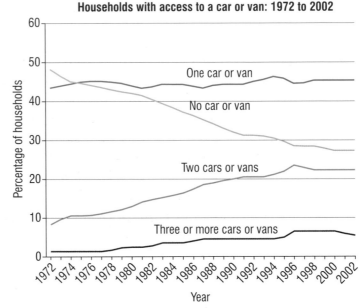

Figure 1 In 1972, about 50% of people had a car they could use. Now around 75% of people have access to a car.

a) Why is there more pollution from cars now than there was 50 years ago?
b) Why is pollution from cars in Moscow as important as pollution from cars in London?

What kinds of pollution?

When we burn any fuel it makes carbon dioxide. Carbon dioxide is a **greenhouse gas**. It collects in the atmosphere and reduces the amount of heat lost by radiation from the surface of the Earth. Most scientists think that this is causing **global warming**, which changes temperatures around the world.

Burning fuels in engines also produce other substances. One group of pollutants is called the ***particulates***. These are tiny particles made up of unburnt hydrocarbons, which scientists think may be especially bad for young children. Particulates may also be bad for the environment too – they travel into the upper atmosphere, reflecting sunlight back into space, causing **global dimming**.

NEXT TIME YOU...

... use a fuel (or electricity which is usually made from natural gas, coal or oil), think how your action affects the environment.

Other pollutants produced by burning fuels include carbon monoxide, sulfur dioxide and nitrogen oxides. Carbon monoxide is formed when there is not enough oxygen present for the fuel to react with oxygen to form carbon dioxide.

Carbon monoxide is a serious pollutant because it affects the amount of oxygen that our blood is able to carry. This is particularly serious for people who have problems with their hearts.

Sulfur dioxide and nitrogen oxides damage us and our environment. In Britain scientists think that the number of people who suffer from asthma and hayfever have increased because of air pollution.

Sulfur dioxide and nitrogen oxides also form acid rain. These gases dissolve in water droplets in the atmosphere and form sulfuric and nitric acids. The rain with a low pH can damage plant and animals.

c) Name four harmful pollutants that may be produced when fossil fuels burn.

Cleaning up our act

We can reduce the effect of burning fossil fuels in several ways. The most obvious way is to remove the pollutants from the gases that are produced when we burn fuels. For some time the exhaust systems of cars have been fitted with *catalytic converters*.

The exhaust gases from the engine travel through the catalytic converter where they pass over transition metals. These are arranged so that they have a very large surface area. This causes the carbon monoxide and nitrogen oxides in the exhaust gases to react. They produce carbon dioxide and nitrogen:

carbon monoxide + nitrogen oxides → carbon dioxide + nitrogen

In power stations, sulfur dioxide is removed from the flue gases by reacting it with quicklime. This is called *flue gas desulfurisation* or FGD for short.

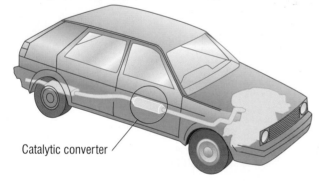

Figure 2 A catalytic converter greatly reduces the carbon monoxide and nitrogen oxides produced by a car engine

Catalytic converter

The methods described here reduce pollution by tackling it after it has been produced. The next page shows how we can also reduce the pollution that fuels produce if we start using alternative fuels.

SUMMARY QUESTIONS

1 Write definitions for the following terms:
   a) greenhouse gas, b) global warming, c) global dimming, d) acid rain

2 A molecule of carbon monoxide requires another atom of oxygen if it is to become a molecule of carbon dioxide. How does a catalytic converter supply this?

3 a) Explain how acid rain is formed and how we are reducing the problem.
   b) Compare the effects of global warming and global dimming.

KEY POINTS

1 Burning fuels releases substances that spread throughout the atmosphere.
2 Some of these substances dissolve in droplets of water in the air, which then fall as acid rain.
3 Carbon dioxide produced from burning fuels is a greenhouse gas. It reduces the rate at which energy is lost from the surface of the Earth by radiation.
4 The pollution produced by burning fuels may be reduced by treating the products of combustion. This can remove substances like nitrogen oxides, sulfur dioxide and carbon monoxide.

C1a 3.5   Alternative fuels

Fuel from plants . . .

The fossil fuels that we use to produce electricity and to drive our cars have some big disadvantages. They produce carbon dioxide, which is a greenhouse gas, and they produce other pollutants too. What's more, once they have all been used there will be no other similar fuels that we can use to replace them – unless we think about the problem and do something about it.

Plants may be one answer to the problem of fuels. For thousands of years people have burned wood to keep themselves warm. Obviously we cannot use wood as a fuel for cars, but there are two ways that plants may be able to keep us on the road. These are explained in Figures 1 and 2.

Figure 1 We can use plants that make sugar to produce ethanol by fermenting the sugar using yeast. We can then add the ethanol to petrol, making **gasohol**. Not only does this reduce the amount of oil needed, it also produces less pollution because gasohol burns more cleanly than pure petrol.

Figure 2 Another new fuel is **biodiesel**. Some plants, like this oilseed rape, produce oils which can be used in diesel engines. We hardly need to make any changes to the engine to do this, and the biodiesel burns very cleanly, like gasohol. These bio-fuels also help tackle global warming. That's because the plants take in carbon dioxide gas during photosynthesis. They still give off carbon dioxide when we burn the bio-fuel – but overall they make little contribution to the greenhouse effect compared with burning fossil fuels.

ACTIVITY

One problem with switching to a new fuel for cars is that people may not trust it, preferring to stick with what they know. One way of convincing them to switch may be to produce advertising material like stickers. People who have switched fuels can then use these on their cars to show other people that the new fuel works just as well as the old fuel.

Design a set of stickers which can be used to make other people think about switching fuels.

... fuel from rubbish

Another way that we could reduce the amount of fuels we use is to replace them with something else – and rubbish seems a good answer! By burning rubbish we could produce some of the energy we need to heat our homes, and we would get rid of a big problem too.

Figure 3 Getting rid of all our rubbish usually means burying it in holes in the ground. This is not a good solution since it is messy, smelly and produces pollution.

Figure 4 We can burn rubbish in an incinerator like this. We can use the energy to heat water which can then heat our homes – or the energy can be used to make electricity.

But producing energy from rubbish is not straightforward. Unless the incinerator is run very carefully, dangerous chemicals called *dioxins* may be produced when the rubbish burns. Although no-one is exactly certain what dioxins do, many people think that they may cause cancer, and that they may damage us in other ways too. So there are arguments on both sides about the benefits of building incinerators.

RECYCLE **NOT** INCINERATE!

BURNING WASTE **CAN** REPLACE **OIL!!**

INCINERATOR WILL **POISON** **OUR** CHILDREN!!

INCINERATOR GOOD FOR ENVIRONMENT AND FOR **LOCAL JOBS!!**

ACTIVITY

You are going to take part in a planning enquiry which will decide whether or not an incinerator should be built. Choose one of the following rôles:

Director of incinerator company
You believe that incinerating waste is the best option – better for the environment (getting rid of waste and producing energy), and better for the local economy. The incinerator would bring real benefits.

Environmental campaigner
You argue that an incinerator is not good. Although the energy produced would replace fossil fuels, you believe that the pollutants the incinerator would produce will be harmful to local people, especially children.

Parent of young children
Although you like the idea of cheap heating and electricity you are concerned that the pollutants, which may be given off by the incinerator, may harm your children.

Local resident
You have lived and worked in the area for years. You feel that the incinerator will bring many benefits to the local area, and that the cheap energy will be a real bonus for local people.

Chair of enquiry
It is your job to manage the enquiry. You will have to give everyone a chance to speak, and you must ensure that everyone gets a fair hearing.

SUMMARY QUESTIONS

1 The following questions are about using hydrocarbons from crude oil as fuels.

a) When hydrocarbon fuels burn in plenty of air, what are the **two** main products?

b) One of these products is particularly bad for the environment – which one, and why?

c) What other substance may be produced when hydrocarbon fuels burn in plenty of air?

d) When hydrocarbon fuels burn inside an engine, what **two** other substances may be produced?

e) What effect do these two substances have on the environment?

f) Diesel engines produce **particulates**. What are they and what effect may they have on the environment?

2 Look at the graph of crude oil prices:

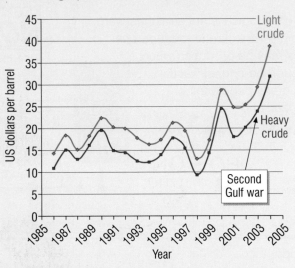

Use the graph to answer the following questions.

a) How do the prices of light crude and heavy crude differ?

b) Why is there this price difference?

c) Why did the price of crude oil rise sharply in 2003?

d) Light crude oil is particularly expensive during summer in the Northern Hemisphere. Suggest **one** possible reason for this.

EXAM-STYLE QUESTIONS

1 The following compounds are found in crude oil:

A C_3H_8 **B** C_8H_{18}

C $C_{12}H_{26}$ **D** $C_{16}H_{34}$

You can use A, B, C or D once, more than once or not at all when answering the questions below.

Which of these compounds

(a) has the highest boiling point?

(b) catches fire most easily?

(c) is collected at the top of the fractionating column when crude oil is distilled?

(d) is the thickest liquid?   (4)

2 Crude oil is a mixture of many different hydrocarbons. Match the words **A**, **B**, **C** and **D** with spaces **1** to **4** in the sentences.

A alkanes   **B** compounds

C fractions   **D** molecules

(a) Crude oil is separated by distillation into ...... containing hydrocarbons with similar boiling points.

(b) Hydrocarbons with the smallest ...... have the lowest boiling points.

(c) Hydrocarbons are ...... of hydrogen and carbon only.

(d) Crude oil contains mostly saturated hydrocarbons called ....   (4)

3 The table shows the number of carbon atoms in the molecules of four fuels obtained from crude oil.

Fuel	Number of carbon atoms in molecules
petroleum gases	2–4
petrol	4–10
kerosene	10–15
diesel oil	14–19

(a) The fuel with the highest boiling point is . . .

     A petroleum gases   **B** petrol

     C kerosene   **D** diesel oil   (1)

(b) Petrol . . .

     A has a higher boiling point than diesel oil.

     B is a thinner liquid than diesel oil.

     C ignites less easily than kerosene.

     D has larger molecules than kerosene.   (1)

(c) The molecule C_4H_{10} could be in . . .

    A petrol only.

    B petrol and kerosene.

    C petrol and petroleum gases.

    D petroleum gases only. (1)

(d) Which one of the following is a saturated hydrocarbon that could be in diesel oil?

    A $C_{12}H_{26}$

    B $C_{16}H_{32}$

    C $C_{17}H_{36}$

    D $C_{18}H_{38}O$ (1)

4 Pentane, C_5H_{12}, is a hydrocarbon fuel. It burns completely in plenty of air.

(a) Name the gas in air that pentane reacts with when it burns. (1)

(b) Write a word equation for the combustion of pentane in plenty of air. (2)

(c) Write a balanced symbol equation for this reaction. (2)

(d) When the air supply is limited a poisonous gas is produced. Name this gas. (1)

(e) Write a balanced symbol equation for the combustion of pentane in a limited supply of air. (2)

5 Suggest two fuels that could be used in place of fossil fuels. Give one advantage and one disadvantage for each of the fuels you have named. (6)

6 Oil companies promote the use of low sulfur fuels.

(a) Explain why it is better to use low sulfur fuels. (3)

(b) Suggest one other reason why oil companies advertise that their fuels are low in sulfur. (1)

7 Crude oil is separated by fractional distillation. In oil refineries this is done in tall towers called fractionating columns.

Give the main steps in this process and explain how the different fractions are separated in a fractionating column. (4)

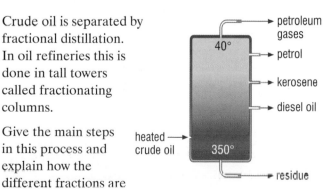

HOW SCIENCE WORKS QUESTIONS

Calculating energy from different fuels

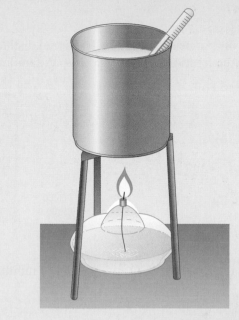

This apparatus can be used to determine the heat given out when different fuels are burned.

The burner is weighed before and after to determine the amount of fuel burned. The temperature of the water is taken before and after, so as to calculate the temperature rise. The investigation was repeated. From this the amount of heat produced by burning a known amount of fuel can be calculated.

a) Construct a table that could be used to collect the data from this experiment. (3)

A processed table of results is given below.

Fuel	Mass burned (g)	Temperature rise (°C)	
Ethanol	4.9	48	47
Propanol	5.1	56	56
Butanol	5.2	68	70
Pentanol	5.1	75	76

b) List three variables that need to be controlled. (3)

c) Describe how you would take the temperature of the water to get the most accurate measurement possible. (2)

d) Do these results show precision? Explain your answer? (2)

e) Would you describe these results as accurate? Explain your answer. (Clue – look at the way the investigation was carried out.) (2)

f) How might you present these results? (1)

EXAMINATION-STYLE QUESTIONS

Chemistry A

See page 57

1   Crude oil can be separated by fractional distillation.
    Match the fractions **A**, **B**, **C** and **D** to the outlets **1** to **4** on the diagram.

   A   diesel oil

   B   kerosene

   C   petrol

   D   petroleum gases

   (diagram: fractional distillation column)
   - 1
   - 40°
   - 2
   - 3
   - 4
   - heated crude oil
   - 350°C
   - residue   *(4 marks)*

See page 30

2   Match the words **A**, **B**, **C** and **D** with the numbers **1** to **4** in the table.

   A   limestone

   B   limewater

   C   quicklime

   D   slaked lime

1	Goes cloudy when reacted with carbon dioxide.
2	Made by thermal decomposition in a kiln
3	Mainly calcium carbonate
4	Solid calcium hydroxide

(4 marks)

See pages 5, 9, 15

3   A student investigated three unknown metal carbonates to compare how easily
    each powder decomposed. She wanted to put the powders in order. She bubbled
    the carbon dioxide gas given off through lime water. She timed how long it took
    for the lime water to look milky. She repeated each test 3 times.
    Here are her results:

Carbonate	Time for lime water to turn milky (s)
A	186, 275, 157
B	90, 163, 142
C	106, 152, 136

   Which statement is true?

   A   Her results were accurate because she repeated them 3 times.

   B   Her conclusion would be reliable and valid.

   C   She collected precise data by repeating her tests.

   D   She could not draw a firm conclusion from the data collected.   *(1 mark)*

GET IT RIGHT!

In multiple choice questions
that ask you to *match* the
letters and numbers, each
letter is used only once and
so if you know three of the
answers, you can answer the
fourth one! In questions with
parts (a) (b) (c) etc. the
letters can be used once,
more than once or not at all
in each question. It is not
worth looking for patterns in
letters for answers because
they are used randomly, so
there is no pattern!

Chemistry B

1 Lead can be extracted from lead oxide using carbon.

See page 39

    (a) Write a word equation for this reaction. *(2 marks)*

    (b) Explain why this is called a reduction. *(1 mark)*

2 Most of the iron produced in a blast furnace is converted into steels.

See page 42

    (a) Why is iron from the blast furnace not very useful? *(1 mark)*

    (b) What are steels? *(2 marks)*

    (c) Explain why steels are harder than pure iron. *(3 marks)*

    (d) Describe how you could compare the strength of an iron wire with a steel wire. You should describe any measurements you would take and how to make it a fair test. *(5 marks)*

See pages 7, 8

3 Complete the equations to show the reactions of limestone and its products.

See pages 29–31

    (a) $CaCO_3 \xrightarrow{\text{heat}} CaO + \ldots\ldots$

    (b) $CaO + \ldots\ldots \longrightarrow Ca(OH)_2$

    (c) $Ca(OH)_2 + \ldots\ldots \longrightarrow CaCO_3 + \ldots\ldots$ *(4 marks)*

4 Read the information in the passage and use it to help you answer these questions.

See page 51

A new method of mining nickel uses plants to extract nickel compounds from the soil. The plants are grown in fields in parts of Canada where there is a higher than usual amount of nickel, but not enough to make it economical to mine normally. The plants are harvested and burnt to produce energy. The ash that is left after burning the crop contains nickel compounds from which the nickel can be extracted. The yield is up to 400 kg of nickel per hectare and farmers are hoping to receive $2000 per hectare for the nickel in the ash from their crops. The current price of nickel is $14 per kg.

    (a) Why is the nickel in the soil not extracted by normal mining? *(1 mark)*

    (b) Explain how the nickel is concentrated in this process so that it can be extracted. *(3 marks)*

    (c) Suggest a method that could be used to produce nickel metal from the ash. *(2 marks)*

    (d) How much are the farmers hoping to receive for each kg of nickel in the ash from their crops? *(1 mark)*

    (e) Why is the amount the farmers hope to receive less than the current price of nickel? *(1 mark)*

GET IT RIGHT!

Always read any information given in a question very carefully. All of the information is there for a reason – you are expected to use it in your answer. Read the questions carefully and refer back to the information before you write your answers. It may help if you highlight or underline key words on your question paper in exams. If you cannot write in this book, make notes on paper as you do the questions. Ask your teacher how best to do this.

C1b | Oils, Earth and atmosphere

What you already know

The face of the Earth has changed over millions of years

Here is a quick reminder of previous work that you will find useful in this unit:

● We can characterise materials by melting point, boiling point and density.

● Mixtures are made up of different substances that are not chemically combined to each other.

● We can separate a mixture of liquids into its different parts using distillation.

● Changes of state involve energy transfers.

● Rocks are formed by processes that take place over different timescales.

● Burning fossil fuels affects our environment.

● We can use indicators to classify solutions as acidic, neutral or alkaline.

RECAP QUESTIONS

1 Where does the water in a puddle go as the puddle dries out when the Sun shines?

2 A car engine will not work properly if there is dirt in the petrol. The dirty petrol is passed through a funnel containing filter paper. How does this clean the petrol?

3 The tea-leaves in a tea bag contain colours and flavours that dissolve in water. Why is it difficult to make iced tea using a tea bag dipped in very cold water?

4 When you put a tray of water in the freezer its temperature drops. At 0°C the water starts to freeze. The temperature then stays the same until all the water is frozen. What is happening?

5 Two fossils are found in two different layers of rocks, one above the other. Which fossil is likely to be the older one – the one found in the upper layer of rock, or the one found in the lower layer? Explain your answer.

6 When we burn fossil fuels, how can it affect the environment?

7 A solution turns universal indicator red – what does this tell you about the solution?

Making connections

Millions of years ago, small plants and animals lived in the oceans of the world. These tiny organisms were supported by energy from the Sun. When they died, the tiny organisms fell to the bottom of the sea. Then they were covered by mud, sand and rock. Over millions of years they became oil . . .

It was only about 100 years ago that people started driving cars! At first only the very rich could afford to buy them. But slowly, as more and more cars were produced, they became less expensive so that more people could own them.

As more and more people bought more and more cars, there were problems with pollution. As well as carbon dioxide causing global warming, car engines produced other pollutants which were bad for people and for other living things too – including plants.

Biodiesel promises an environmentally friendly fuel which may one day replace fossil fuels. Made from plant oils, biodiesel has much less effect on the environment than ordinary diesel fuel.

ACTIVITY

Imagine that you are going to present a short radio programme about the effect that the motor car has had on our lives. Using the information here, write and present this programme to other people in your class.

Chapters in this unit

Products from oil   Plant oils   The changing world

C1b 4.1   Cracking hydrocarbons

LEARNING OBJECTIVES

1 How do we make smaller, more useful molecules from crude oil?
2 What are alkenes and how are they different from alkanes?

Many of the fractions that we get by distilling crude oil are not very useful. They contain hydrocarbon molecules which are too long for us to use them as fuels that are in high demand. The hydrocarbons that contain very big molecules are thick liquids or solids with high boiling points. They are difficult to vaporise and do not burn easily – so they are no good as fuels! Yet the main demand from crude oil is for fuels.

Figure 1 Huge crackers like this are used to split large hydrocarbon molecules into smaller ones

Luckily we can break down large hydrocarbon molecules in a process we call **cracking**. The best way of breaking them up uses heat and a catalyst, so we call this *catalytic cracking*. The process takes place in a *cat cracker*.

In the cracker a heavy fraction produced from crude oil is heated strongly to turn the hydrocarbons into a gas. This is passed over a hot catalyst where thermal decomposition reactions take place. The large molecules split apart to form smaller, more useful ones.

a) Why is cracking so important?
b) How are large hydrocarbon molecules cracked?

Example of cracking

Decane is a medium sized molecule with ten carbon atoms. When we heat it to 800°C with a catalyst it breaks down. One of the molecules produced is pentane which is used in petrol. We also get propene and ethene which we can use to produce other chemicals.

$$C_{10}H_{22} \xrightarrow{\text{800°C + catalyst}} C_5H_{12} + C_3H_6 + C_2H_4$$
decane   pentane   propene   ethene

This reaction is an example of **thermal decomposition**.

Notice how this cracking reaction produces different types of molecules. One of the molecules is pentane. The first part of its name tells us that it has five carbon atoms (*pent-*). The last part of its name (*-ane*) shows that it is an alkane. Like all other alkanes, pentane is a **saturated hydrocarbon** – its molecule has as much hydrogen as possible in it.

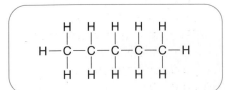

Figure 2 A molecule of pentane

The other molecules in this reaction have names that end slightly differently. They end in *-ene*. We call this type of molecule an **alkene**. The different ending tells us that these molecules are **unsaturated** because they contain a *double bond* between two of their carbon atoms. Look at Figure 3:

Alkenes with one double bond have this general formula, C_nH_{2n}.

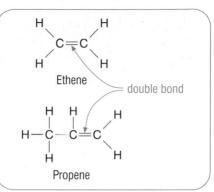

Figure 3 A molecule of propene and a molecule of ethene. These are both alkenes – each molecule has a carbon–carbon double bond in it.

PRACTICAL

Cracking

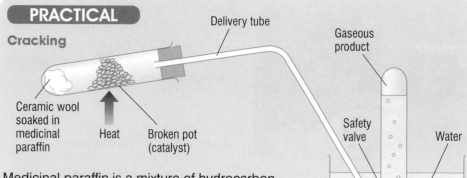

Ceramic wool soaked in medicinal paraffin

Heat

Broken pot (catalyst)

Delivery tube

Gaseous product

Safety valve

Water

Medicinal paraffin is a mixture of hydrocarbon molecules. You can crack it by heating it and passing the vapour over hot pieces of broken pot. The products that you make in this reaction are insoluble gases, so you can collect them by bubbling them through water.

If you carry out this practical, collect at least two test tubes of gas. Test one by putting a lighted splint into it. Test the other by adding a few drops of bromine water to it.

● Why must you remove the end of the delivery tube from the water when you stop heating?

A simple experiment like the one described above shows that alkenes burn. They also react with bromine water (which is orange-yellow) – the products of this reaction are colourless. This means that we have a good test to see if a hydrocarbon is unsaturated:

unsaturated hydrocarbon + bromine water → products
    (colourless)   (orange-yellow)   (colourless)

saturated hydrocarbon + bromine water → no reaction
    (colourless)   (orange-yellow)   (orange-yellow)

DID YOU KNOW...

... that ethene is a really important chemical? Although we know over 500 ways of making ethene, the only way that it is currently made commercially is from oil.

SUMMARY QUESTIONS

1 Copy and complete using the words below:

alkenes   catalyst   cracking   double   heating   unsaturated

We can break down large hydrocarbon molecules by ...... them and passing them over a .... This is called .... Some of the molecules produced when we do this contain a ...... bond – they are called ...... hydrocarbons and we call them ....

2 Cracking a hydrocarbon makes two new hydrocarbons, A and B. When bromine water is added to A, nothing happens. Bromine water added to B loses its colour. Which hydrocarbon is unsaturated?

3 a) An alkene molecule with one double bond contains 7 carbon atoms. How many hydrogen atoms does it have? Write down its formula.

   b) Decane (with 10 carbon atoms) is cracked into octane (with 8 carbon atoms) and ethene. Write a balanced equation for this reaction.

KEY POINTS

1 We can split large hydrocarbon molecules up into smaller molecules by heating them and passing the gas over a catalyst,

2 Cracking produces unsaturated hydrocarbons, which we call alkenes.

3 Alkenes burn, and also react with bromine water, producing colourless products.

C1b 4.2   Making polymers from alkenes

Refining crude oil produces a huge range of hydrocarbon molecules which are very important to our way of life. Oil products are all around us. We simply cannot imagine life without them.

Figure 1 All of these products were manufactured using chemicals made from oil

The most obvious way that we use hydrocarbons from crude oil is as fuels. We use fuels in our transport and at home. We also use them to make electricity in oil-fired power stations.

Then there are the chemicals we make from crude oil. We use them to make things ranging from margarines to medicines, from dyes to explosives. But one of the most important ways that we use chemicals from oil is to make plastics.

Plastics

Plastics are made up of huge molecules made from lots of small molecules that have joined together. We call the small molecules **monomers**. We call the huge molecules they make **polymers** – *mono* means 'one' and *poly* means 'many'. We can make different types of plastic which have very different properties by using different monomers.

a) List three ways that we use fuels.
b) What are the small molecules that make up a polymer called?

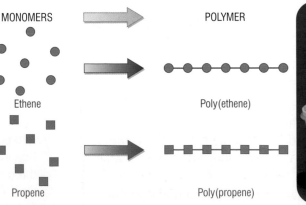

Figure 2 Polymers produced from oil are all around us and are part of our everyday lives

Ethene (C_2H_4) is the smallest unsaturated hydrocarbon molecule. We can turn it into a polymer known as poly(ethene) or polythene. Polythene is a really useful plastic. It is easy to shape, strong and transparent (unless we add colouring material to it). 'Plastic' bags, plastic drink bottles, dustbins and clingfilm are all examples of polythene that are very familiar to us in everyday life.

Propene (C_3H_6) is another alkene. We can also make polymers with propene as the monomer. The plastic formed is called poly(propene) or polypropylene. Poly(propene) is a very strong, tough plastic. We can use it to make many things, including milk crates and ropes.

c) Is ethene an alkane or an alkene?

d) Which plastic can we make from the monomer called propene?

How do monomers join together?

When alkene molecules join together, the double bond between the carbon atoms in each molecule 'opens up'. It is replaced by single bonds as thousands of molecules join together. This is an example of an *addition reaction*. Because a polymer is made, we call it *addition polymerisation*.

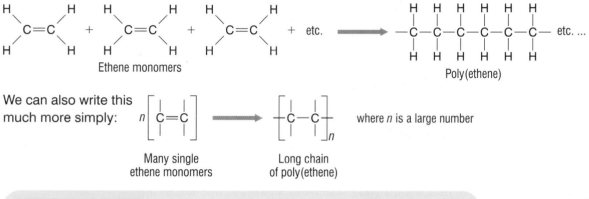

Ethene monomers · · · · · · · · · · · · · · Poly(ethene)

We can also write this much more simply: where n is a large number

Many single ethene monomers · · · · · · · · Long chain of poly(ethene)

PRACTICAL

Modelling polymerisation

Use a molecular model kit to show the polymerisation of ethene to form poly(ethene).

Make sure you can see how the equation shown above represents the polymerisation reaction you have modelled.

● Describe what happens to the bonds in the reaction.

GET IT RIGHT!

The double C=C bond in ethene (an alkene) makes it much more reactive than ethane (an alkane).

e) Think up a model to demonstrate the polymerisation of ethene, using people in your class as monomers. Evaluate the ideas of other groups.

SUMMARY QUESTIONS

1 Copy and complete using the words below:

addition   ethene   monomers   polymers

Plastics are made out of large molecules called .... We make these by joining together lots of small molecules called .... One example of a plastic is poly(ethene), made from .... Poly(ethene) is formed as a result of an ...... reaction.

2 Why is ethene the smallest unsaturated hydrocarbon molecule?

3 a) Draw a propene molecule.
   b) Draw structures to show how propene molecules join together to form poly(propene).
   c) Explain the polymerisation reaction in b).

KEY POINTS

1 Plastics are made of polymers.
2 Polymers are large molecules made when monomers (small molecules) join together.

C1b 4.3   The properties of plastics

As you have just seen on pages 72 and 73, we can make plastics from chemicals made from crude oil. Small molecules called monomers join together to make much bigger molecules called polymers. As the monomers join together they produce a tangled web of very long chain molecules.

The atoms in these chains are held together by very strong chemical bonds. This is true for all plastics. But the size of the forces *between* polymer molecules in different plastics can be very different.

We call the forces between molecules **intermolecular forces**. The size of the intermolecular forces between the polymer molecules in a plastic depends partly on:

● the monomer used, and
● the conditions we choose to carry out polymerisation.

a) How are the atoms in polymer chains held together?
b) What do we call the forces between polymer chains?

In some plastics the forces between the polymer molecules are weak. When we heat the plastic, these weak intermolecular forces are broken and the plastic becomes soft. When we cool the plastic, the intermolecular forces bring the polymer molecules back together, and the plastic hardens again. We call plastics which behave like this **thermosoftening** plastics.

Poly(ethene), poly(propene) and poly(chloroethene) are all examples of thermosoftening plastics.

Figure 1 The forces between the molecules in poly(ethene) are relatively weak. This means that this plastic softens fairly easily when heated.

Figure 2 We usually call poly(chloroethene) by its more common name – polyvinylchloride, or PVC for short. We make lots of everyday plastic articles from PVC.

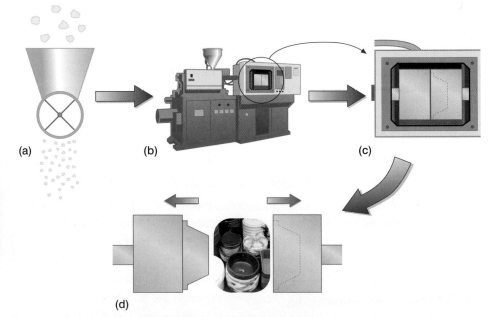

(a)   (b)   (c)

(d)

Figure 3 A very common way of making things out of polymers is to use a thermosoftening plastic that can be shaped in a mould: **(a)** chunks of monomer are ground into small pieces. **(b)** These are heated to melt them and then . . . **(c)** the molten plastic is forced into a mould. **(d)** The mould is separated to release the finished article.

There are two types of poly(ethene) – high density (HDPE) and low density poly(ethene) (LDPE). Both are made from ethene monomers but are formed under different conditions.

Using very high pressures and a trace of oxygen, ethene forms LDPE. The polymer chains are branched and they can't pack closely together.

Using a catalyst at 50°C and a slightly raised pressure gives us HDPE. This is made up of straight poly(ethene) molecules. They can pack closely together. Therefore forces between molecules (intermolecular forces) are stronger. HDPE has a higher softening temperature and is stronger than LDPE.

We cannot soften all plastics. Some monomers also make chemical bonds between the polymer chains when they are first heated in order to shape them. These bonds are strong, and they stop the plastic from softening when we heat it in the future. We call this type of polymer a **thermosetting plastic**.

c) What do we call a plastic that softens when we heat it?
d) What do we call a plastic that does not soften again once it has been made?

Figure 4 Plastic kettles are made out of thermosetting plastics

PRACTICAL

Modifying a polymer

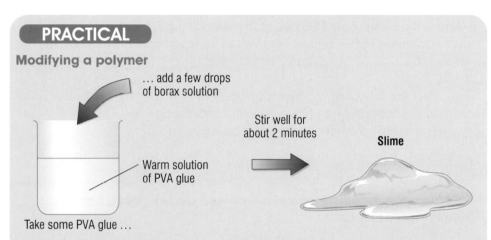

... add a few drops of borax solution

Stir well for about 2 minutes

Slime

Warm solution of PVA glue

Take some PVA glue ...

● How could you investigate if the properties of slime depend on how much borax you add?

The glue becomes slimy because the borax makes the long polymer chains in the glue link together to form a jelly-like substance.

SUMMARY QUESTIONS

1 Copy and complete using the words below:

    bonds   thermosetting   tangled   weak

The polymer chains in a plastic form a ...... web. In the chain, the atoms are held together by strong .... If the intermolecular forces between the polymer chains are ...... , the plastic softens at a relatively low temperature. Some polymers have strong bonds between the chains – we call these ...... polymers.

2 Why do we use thermosetting plastics to make plastic kettles?

3 Polymer A starts to soften at 400°C while polymer B softens at 150°C.

Explain this statement using ideas about intermolecular forces.

KEY POINTS

1 Monomers affect the properties of the polymers that they produce.
2 Changing reaction conditions can also change the type of polymer that is produced.

C1b 4.4   New and useful polymers

Figure 1 Plastic drinks bottles are made from a plastic called poly(etheneterephthalate), or PET for short

You have probably heard the question – 'which came first, the chicken or the egg?' Polymers and the way that we use them are a bit like this. That's because sometimes we use a polymer to do a job because of its properties. But then at other times we might design a polymer with special properties so that it can do a particular job.

The bottles that we buy fizzy drinks in are a good example of a polymer that we use because of its properties. These bottles are made out of a plastic called PET.

The polymer it is made from is ideal for making drinks bottles. It produces a plastic that is very strong and tough, and which can be made transparent. The bottles made from this plastic are much lighter than glass bottles. This means that they cost less to transport.

a) Why is the plastic called PET used to make drinks bottles?
b) Why do drinks in PET bottles cost less to transport than drinks in glass bottles?

Now, rather than choosing a polymer because of its properties, materials scientists are designing new polymers with special properties. These are polymers that have the right properties to do a certain job.

c) What do we mean by a 'designer polymer'?

Medicine is one area where we are beginning to see big benefits from these 'polymers made to order'.

Figure 2 A sticking plaster is often needed when we cut ourselves. Getting hurt isn't much fun – and sometimes taking the plaster off can be painful too.

We all know how uncomfortable pulling a plaster off your skin can be. But for some of us taking off a plaster is really painful. Both very old and very young people have quite fragile skin. But now a group of chemists has made a plaster where the 'stickiness' can be switched off before the plaster is removed. The plaster uses a light sensitive polymer. Look at Figure 3.

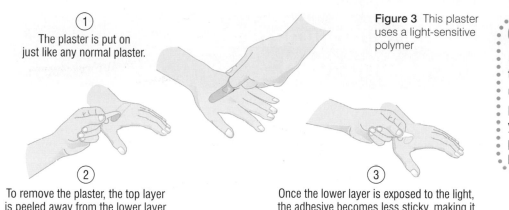

Figure 3 This plaster uses a light-sensitive polymer

① The plaster is put on just like any normal plaster.

② To remove the plaster, the top layer is peeled away from the lower layer which stays stuck to the skin.

③ Once the lower layer is exposed to the light, the adhesive becomes less sticky, making it easy to peel off your skin.

NEXT TIME YOU...

. . . take a plaster off your skin, think about the technologies used to create it. Did it hurt to pull it off? Did it leave a mark on your skin? Are there ways the plaster could be made even better?

New polymers can also come to our rescue when we are cut badly enough to need stitches. A new shape memory polymer is being developed by doctors which will make stitches that keep the sides of a cut together. Not only that, but the polymer will also dissolve once it has done its job. So there will be no need to go back to the doctor to have the stitches out.

Figure 4 When a shape memory polymer is used to stitch a wound loosely, the temperature of the body makes the thread tighten and close the wound, applying just the right amount of force. This is an example of a '**smart polymer**' i.e. one that changes in response to changes around it. In this case a change in temperature causes the polymer to change its shape. Later, after the wound is healed, the material is designed to dissolve and is harmlessly absorbed by the body.

PRACTICAL

Evaluating plastics

Carry out an investigation to compare the suitability of different plastics for a particular use.

For example, you might look at treated and untreated fabrics for waterproofing and 'breatheability' (gas permeability) or different types of packaging.

SUMMARY QUESTIONS

1 Copy and complete using the words below:

cold   hot   PET   properties   shape   strong   transparent

We choose a polymer for a job because it has certain .... For example, we make drinks bottles out of a plastic called ...... because it is ...... and ....
Scientists can also design 'smart' polymers, for example ...... memory polymers. These change their shape when they are ...... or ....

2 a) The polymer in some sticking plasters is switched off by light because light makes bonds form between the polymer chains. Suggest why this may make the polymer less sticky.
   b) Design a leaflet for a doctor to give to a patient, explaining how stitches made from smart polymers work.

KEY POINTS

1 New polymers are being developed all the time. They are designed to have properties that make them specially suited for certain uses.
2 Smart polymers may have their properties changed by light, temperature or by other changes in their surroundings.

C1b 4.5

Plastics, polymers and packaging food

Shopping for food

> Buying things in a shop like this probably took longer than shopping in the supermarket. But buying food already packed isn't just quicker – the packaging material used to wrap the food is designed to keep it fresh for longer, and to help to stop dirt and germs getting into it. And new packaging materials that we are developing will do this job even better – for example, they'll allow air in and out of the package without letting the food dry out.

> Look at this shop – it's a bit different to the supermarket where we do our shopping each week! For a start, everything needs weighing out before you can buy it. Shopping like this must have taken ages!

> Yeah – but my Gran says her Mum used to go shopping nearly every day, not just once a week! My Mum buys everything already wrapped up from the supermarket – except cheese. She always gets that from the deli at the supermarket where they weigh it out for you – I don't know why.

ACTIVITY

To wrap or not to wrap?

You are going to think about whether we should buy food that has already been wrapped – or whether it would be better if we had it weighed and wrapped when we bought it.

Start off by working on your own. Draw up a table like the one below, and use it to list reasons for buying food already wrapped and for buying food that needs to be wrapped.

When you have drawn up your lists, compare them with someone else in your class. Discuss any differences in your tables.

Buying food that has not been wrapped	Buying food that is already wrapped

'Would you like that in a bag, sir?'

When you buy something in a shop, you need to get it home safely. If you're a well-organised kind of person you may have taken a bag with you – but most of us are quite happy to say 'yes' when the shop assistant asks us 'Would you like a bag for that?'And when we get home – what happens to the bag?

So just how many bags do we use in one year? How many bags does one supermarket give out in a year? Is this a problem?

Are the bags **biodegradable** (will they be broken down in nature by microorganisms when we throw them away)? Or will they take up valuable space in land-fill sites for years to come?

ACTIVITY

How many bags?

You are going to carry out a survey to find out how many carrier bags people use.
Design a questionnaire to answer questions like:

- how many carrier bags do you have at home?
- how many carrier bags do you bring home from the supermarket each week?
- do you use re-usable bags for your shopping?
- what happens to the carrier bags once you have used them?

When you have collected the answers to your questions, present the information using tables, graphs and charts.
Discuss the questions below:

- What do you think about the number of carrier bags used by people? Should we try to reduce this? Why (not)? How?
- Is your data representative of your local area? How could you be more certain?

What factors, if any, might affect the number of carrier bags used in different parts of the country?
How could you collect data to best reflect the picture in the whole country?

SUMMARY QUESTIONS

1 Write simple definitions for the following words:

   a) cracking

   b) distillation

   c) refining

   d) saturated hydrocarbon

   e) unsaturated hydrocarbon.

2 Propene is a hydrocarbon molecule containing three carbon atoms and six hydrogen atoms.

   a) What is meant by the term **hydrocarbon**?

   b) Draw the structural formula of propene.

   c) Is propene a **saturated** molecule or an **unsaturated** molecule? Explain your answer.

   d) Test tubes A and B in the diagram look identical. One test tube contains propane, while the other test tube contains propene.

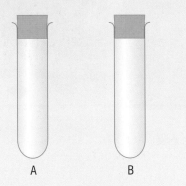

A   B

   Explain clearly what test you could carry out to find out which tube contains propene, stating clearly the results obtained in each case.

   e) Propene molecules will react together to form long chains.
   i) What do we call this type of reaction?
   ii) What properties does this new substance have?
   iii) Why does the widespread use of substances like this present an environmental problem?

3 a) What is a **polymer**?

   b) How do **monomers** join together to make a polymer?

   c) What factors can affect the way that a polymer behaves?

4 'Of all the problems facing us when disposing of our waste, the problem of plastics is the greatest.' Write a paragraph setting out arguments both agreeing and disagreeing with this statement.

EXAM-STYLE QUESTIONS

1 This question is about polymers.
   Match the words **A**, **B**, **C** and **D** with the spaces **1** to **4** in the sentences.

   A transparent

   B light sensitive

   C thermosetting

   D thermosoftening

   A plastic that can be remoulded is made from a ......**1**...... polymer.

   The plastic used to make a handle for a grill pan would be best made from a polymer that is ......**2**......

   The polymers used to make food wrappings for use in a supermarket are best if they are ......**3**......

   Some new types of sticking plaster can be removed easily because they are made from ......**4**......polymers.
   (4)

2 Match the words **A**, **B**, **C** and **D** with the descriptions **1** to **4** in the table.

   A bromine

   B butene

   C poly(ethene)

   D propane

	Description
1	An alkane
2	A polymer
3	Used to make polymers
4	Used to test for unsaturation

   (4)

3 Match the words **A**, **B**, **C** and **D** with the spaces **1** to **4** in the sentences about plastics.

   A chemical bonds

   B intermolecular forces

   C monomers

   D polymers

   Plastics are made of many ......**1**...... joined together in long chains to form ......**2**...... . In thermosoftening plastics the long chains are held together by ......**3**...... which are overcome by heating. In thermosetting plastics ......**4**...... form between the long chains when they are heated and the plastic sets hard.
   (4)

4 An alkane, $C_{12}H_{26}$, was cracked. The reaction that took place is represented by the equation:

$$C_{12}H_{26} \rightarrow C_6H_{14} + C_4H_8 + \ldots\ldots$$

(a) The formula of the missing compound is …

    A CH_4 **B** C_2H_4

    C C_2H_6 **D** C_3H_8

(b) The compound C_4H_8 is …

    A an alkane   **B** an alkene

    C a poly(alkane)   **D** a poly(alkene)

(c) Addition polymers can be made from …

    A $C_{12}H_{26}$ **B** C_6H_{14}

    C C_4H_8 **D** C_2H_6

(d) The structure of C_4H_8 could be …

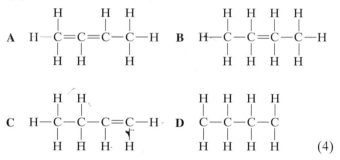

(4)

5 Read the passage about 'Slime' and use the information to help you answer the questions.

'Slime' has some of the properties of a liquid and some of the properties of a solid. It can be poured but it bounces if dropped on the floor. 'Slime' is made by mixing a solution containing a polymer called PVA with borax. When the substances are mixed the borax forms cross-links between the polymer chains. Some of the cross-links are chemical bonds and some are intermolecular forces involving water molecules. Lots of water molecules are held between the polymer chains and these give 'Slime' its flexibility and fluidity.

(a) Describe a molecule of a typical polymer. (2)

(b) Suggest why 'Slime' has the properties of both a solid and a liquid. (3)

(c) Suggest one method that you could use to modify the properties of 'Slime'. (1)

(d) A student tested different types of 'Slime' by measuring how far they stretched before they broke.

    (i) What was the independent variable in the investigation? (1)

    (ii) What type of variable was the dependent variable – categoric, ordered, discrete or continuous? (1)

HOW SCIENCE WORKS QUESTIONS

Biodegradable plastics could be used for growing crops

Non-biodegradable plastic has been used for many years for growing melons. The plants are put into holes in the plastic and their shoots grow up above the plastic. The melons are protected from the soil by the plastic and grow with very few marks on them. Biodegradable plastic has been tested – to reduce the amount of non-recycled waste plastic.

In this investigation two large plots were grown. One using biodegradable plastic, the other using normal plastic. The results were as follows:

Plastic used	Early yield (kg/hectare)	Total yield (kg/hectare)	Average melon weight (kg)
Normal	210	4 829	2.4
Biodegradable	380	3 560	2.2

a) This was a field investigation. Describe how the experimenter would have chosen the two plots. (3)

b) The hypothesis was that the biodegradable plastic would produce less fruit than the normal plastic. Is the hypothesis supported or refuted, or should a new hypothesis be considered? Explain your answer. (2)

c) How could the accuracy of this investigation be improved? (1)

d) How could the reliability of these results be tested? (1)

e) How would you view these results if you were told that they were funded by the manufacturer of the normal plastic? (1)

C1b 5.1   Extracting vegetable oils

LEARNING OBJECTIVES

1 How do we extract oils from plants?
2 Why are vegetable oils important foods?
3 What are unsaturated oils and how do we detect them?

Plants use the Sun's energy to produce glucose from carbon dioxide and water:

$$6CO_2 + 6H_2O \xrightarrow{\text{energy (from sunlight)}} C_6H_{12}O_6 + 6O_2$$

Plants then turn glucose into other chemicals using more chemical reactions. In some cases these other chemicals can be very useful to us – for example the **vegetable oils** that some plants make.

Farmers plant crops like oilseed rape, which is an important source of vegetable oil. We find the oils in the seeds of the rape plant. Once it has flowered and set seed, the farmer collects the seeds using a combine harvester.

The seeds are then taken to a factory where they are crushed, then pressed to extract the oil in them. The impurities are removed from the oil, and we can process it to make it into important foods, as we shall see later in this chapter.

Figure 1 Oilseed rape is a common sight in our countryside. As its name tells us, it is a good source of vegetable oil.

We extract other vegetable oils using steam. For example, we can extract lavender oil from lavender plants by distillation. The plants are put into boiling water and the oil from the plant evaporates. It is then collected by condensing it, when the water and other impurities can be removed to give pure lavender oil.

Figure 2 Norfolk lavender oil is extracted from lavender plants using steam

PRACTICAL

Extracting plant oil by distillation

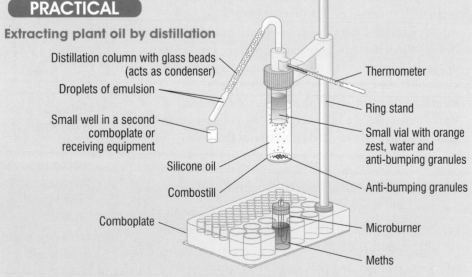

Distillation column with glass beads (acts as condenser)
Droplets of emulsion
Small well in a second comboplate or receiving equipment
Silicone oil
Combostill
Comboplate
Thermometer
Ring stand
Small vial with orange zest, water and anti-bumping granules
Anti-bumping granules
Microburner
Meths

Take care not to let the contents of the small vial boil over.

● What does the liquid collected look and smell like?

a) Write down *two* ways that can be used to extract vegetable oils from plants.

Vegetable oils as foods

Vegetable oils are very important foods because they contain a great deal of energy, as the table shows.

There are lots of different vegetable oils. Each vegetable oil has slightly different molecules. However, all vegetable oils have molecules which contain chains of carbon atoms with hydrogen atoms attached to them:

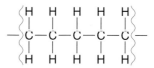

In some vegetable oils there are carbon atoms joined to each other by carbon–carbon double bonds (C=C). We call these **unsaturated oils**. We can detect the double bonds in unsaturated oils with bromine solution, as you did with the double bonds in alkene molecules. (See page 71.) They also react with iodine solution.

This provides us with an important way of detecting unsaturated oils:

unsaturated oil + bromine solution (yellow/orange) → colourless solution
unsaturated oil + iodine solution (violet/reddy brown depending on solvent) → colourless solution

b) What will you see if you test a polyunsaturated margarine with iodine water?

Energy in vegetable oil and other foods	
Food	**Energy in 100 g (kJ)**
vegetable oil	3900
sugar	1700
animal protein (meat)	1100

Figure 3 Vegetable oils have a high energy content

PRACTICAL

Testing for unsaturation

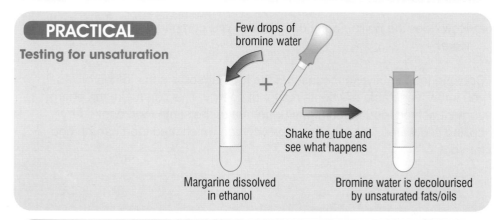

Few drops of bromine water

Shake the tube and see what happens

Margarine dissolved in ethanol

Bromine water is decolourised by unsaturated fats/oils

DID YOU KNOW...

. . . that no more than 20% of the energy in your diet should come from fats?

SUMMARY QUESTIONS

1 Copy and complete using the words below:

bromine distillation energy iodine pressing unsaturated

We can extract vegetable oils from some plants by ...... or .... Vegetable oils are particularly important as foods because they contain a lot of .... Some vegetable oils contain carbon–carbon double bonds – we call these ...... vegetable oils. They can be detected by reacting them with ...... or ...... solution.

2 Why might a diet containing too much vegetable oil make you fat?

3 A sample of vegetable oils is tested with iodine solution. The solution is decolourized. Which of the following statements is true?

a) The sample contains *only* unsaturated oils.
b) The sample contains *only* saturated oils.
c) The sample may contain a mixture of saturated and unsaturated oils.

Explain your answer.

KEY POINTS

1 Vegetable oils can be extracted from plants by pressing or by distillation.
2 Vegetable oils are important foods.
3 Unsaturated oils contain carbon–carbon double bonds. We can detect them using bromine or iodine solutions.

C1b 5.2   Cooking with vegetable oils

The temperature that a liquid boils at depends on the size of the forces between its molecules. The bigger these forces are, the higher the liquid's boiling point.

The molecules in vegetable oils are much bigger than water molecules. This makes the forces between the molecules in vegetable oils much larger. So the boiling points of vegetable oils are much higher than the boiling point of water.

When we cook food we heat it to a temperature where chemical reactions cause permanent changes to happen to the food. When we cook food in vegetable oil the result is very different to when we cook it in water. This is because vegetable oils can be used at a much higher temperature than boiling water.

So the chemical reactions that take place are very different. The food cooks more quickly, and very often the outside of the food turns a different colour, and becomes crisper.

a) How does the boiling point of vegetable oils compare to the boiling point of water?

Cooking food in oil also means that the food absorbs some of the oil. As you know, vegetable oils contain a lot of energy. This can make the energy content of fried food much higher than that of the same food cooked by boiling it in water. This is one reason why too much fried food can be bad for you!

Figure 1 An electric fryer like this one enables vegetable oil to be heated safely to a high temperature.

Figure 2 Boiled potatoes and fried potatoes are very different. One thing that probably makes chips so tasty is the contrast of crisp outside and soft inside, together with the different taste produced by cooking at a higher temperature. The different colour may be important too.

PRACTICAL

Investigating cooking

Compare the texture and appearance of potato pieces after equal cooking times in water and oil.

You might also compare the cooking times for boiling, frying and oven baking chips.

If possible carry out some taste tests in hygienic conditions.

GET IT RIGHT!

No chemical bonds are broken when vegetable oils melt or boil – these are physical changes.
When oils are hardened with hydrogen, a chemical change takes place, producing margarine (which has a higher melting point than the original oil).

b) How is food cooked in oil different to food cooked in water?

Unsaturated vegetable oils are usually liquids at room temperature. This is because the carbon–carbon double bonds in their molecules stop the molecules fitting together very well. This reduces the size of the forces between the molecules.

The boiling and melting points of these oils can be increased by adding hydrogen to the molecules. The reaction replaces some or all of the carbon–carbon double bonds with carbon–carbon single bonds. This allows the molecules to fit next to each other better. So the size of the forces between the molecules increases and the melting point is raised.

With this higher melting point, the liquid oil becomes a solid at room temperature. We call changing a vegetable oil like this **hardening** it.

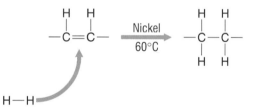

Figure 3 We harden a vegetable oil by reacting it with hydrogen. To make the reaction happen, we must use a nickel catalyst, and carry it out at about 60°C.

c) What do we call it when we add hydrogen to a vegetable oil?

Oils that we have treated like this are sometimes called *hydrogenated oils*. Because they are solids at room temperature, it means that they can be made into spreads to be put on bread. We can also use them to make cakes, biscuits and pastry.

Figure 4 We can use hydrogenated vegetable oils in cooking to make a huge number of different, and delicious, foods!

SUMMARY QUESTIONS

1 Copy and complete using the words below:

energy   hardening   higher   hydrogen   melting   tastes

The boiling points of vegetable oils are ...... than the boiling point of water. This means that food cooked in oil ...... different to boiled food. It also contains more ....
The boiling and ...... points of oils may also be raised by adding ...... to their molecules. We call this ...... the oil.

2 a) Why are hydrogenated vegetable oils more useful than oils that have not been hydrogenated?
   b) Explain how we harden vegetable oils and why the melting point is raised.

KEY POINTS

1 Vegetable oils are useful in cooking because of their high boiling points.
2 Vegetable oils are hardened by reacting them with hydrogen to increase their boiling and melting points.

C1b 5.3   Everyday emulsions

1 What are emulsions and how do we make them?
2 Why are emulsions made from vegetable oils so important?

Figure 1 Smooth food has a good texture and looks as if it will taste nice – but it is not always easy to make, or to keep it smooth

The texture of food – what it feels like in your mouth – is a very important part of foods. Smooth foods like ice cream and mayonnaise are examples of food that might seem simple. In fact, they are quite complicated and difficult to make.

Smooth foods are made from a mixture of oil and water. Everyone knows that oil and water don't mix. Just try it by pouring a little cooking oil into a glass of water.

But we can persuade these two very different substances to mix together by making the oil into very small droplets. These spread out throughout the water and produce a mixture called an **emulsion**.

A good example of this is milk. Milk is basically made up of small droplets of animal fat dispersed in water.

Figure 2 Milk is an emulsion made up of animal fat and water, together with some other substances

PRACTICAL

A closer look at milk

You can view different types of milk – skimmed, semi-skimmed and full fat – under a microscope.

● What differences do you think you will see? Try it out.

Emulsions often behave very differently to the things that we make them from. For example, mayonnaise is made from ingredients that include oil and water. Both of these are runny – but mayonnaise is not!

Another very important ingredient in mayonnaise is egg yolks. Apart from adding a nice yellow colour, egg yolks have a very important job to do in mayonnaise. They stop the oil and water from separating out into layers. Food scientists have a word for this type of substance – they call them **emulsifiers**.

a) What do we mean by 'an emulsifier'?

Emulsifiers make sure that the oil and water in an emulsion cannot separate out. This means that the emulsion stays thick and smooth. Any creamy sauce needs an emulsifier. Without it we would soon find blobs of oil or fat floating around in the sauce. This might make us think that something had gone wrong in the cooking process. It may even make you feel quite sick!

b) How does an emulsifier help to make a good creamy sauce?

One very popular emulsion is *ice cream*. Everyday ice cream is usually made from vegetable oils, although luxury ice cream may also use animal fats.

Ice cream is one of the most complicated foods to make. To make the ice cream light, a great deal of air needs to be beaten into it to make a *foam*. The air is held in this foam by adding **stabilisers** and **thickeners**. These are substances with molecules that produce large 'cages' full of air when they are mixed with water.

Emulsifiers then keep the oil and water mixed together in the ice cream while we freeze it. Without them, the water in the ice cream freezes separately, producing crystals of ice. It makes the ice cream crunchy rather than smooth. This happens if you allow ice cream to melt and then put it back in the freezer.

Figure 3 Ice cream is a very complicated mixture of chemicals

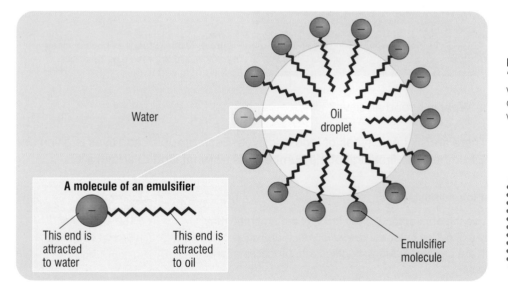

Water

Oil droplet

A molecule of an emulsifier

This end is attracted to water

This end is attracted to oil

Emulsifier molecule

Figure 4 An emulsifier is a molecule with 'a tail' that 'likes' oil and 'a head' that 'likes' water. The charged oil droplets repel each other, keeping them spread throughout the water.

SUMMARY QUESTIONS

1 Copy and complete using the words below:

> emulsifier   emulsion   ice cream   mayonnaise
> mix   separating   small

Oil and water do not ...... together, but if the oil droplets can be made very ...... it is possible to produce a mixture of oil and water called an .... To keep the oil and water from ...... we can use a chemical called an .... Important examples of food made like this include ...... and ....

2 Salad cream is an emulsion made from vegetable oil and water. In what ways is salad cream different from both oil and water?

3 a) Why do we need to add an emulsifier to an emulsion like salad cream?
   b) Explain how emulsifier molecules do their job.

KEY POINTS

1 Oils do not dissolve in water.
2 Oils can be used to produce emulsions which have special properties.
3 Emulsions made from vegetable oils are used in many foods, such as salad dressings and ice creams.

C1b 5.4   What is added to our food?

LEARNING OBJECTIVES

1 What are food additives and why are they put in our food?
2 How can we detect food additives?

People have always needed to find ways of making food last longer – to *preserve* it. For hundreds of years we have added substances like salt or vinegar to food in order to keep it longer. As our knowledge of chemistry has increased we have used other substances too, to make food look or taste better.

We call a substance that is added to food to make it keep longer or to improve its taste or appearance a **food additive**. Additives that have been approved for use in Europe are given **E numbers** which identify them. For example E102 is a yellow food colouring called *tartrazine*, while E220 is the preservative *sulfur dioxide*.

Figure 1 Modern foods contain a variety of additives to improve their taste or appearance, and to make them keep longer

DID YOU KNOW...

... that lemon juice can prevent a cut apple turning brown, thanks to the antioxidant action of vitamin C?

a) What is a food additive?

There are six basic types of food additives. Each group of additives is given an E number. The first digit of the number tells us what kind of additive it is.

E number	Additive	What the additive does	Example
E1 . . .	colours	Improve the appearance of the food. There are three main classes of colour in foods: *natural colours*, *browning colours*, which are produced during cooking and processing, and *additives*.	E150 – caramel, a brown colouring
E2 . . .	preservatives	Help food to keep longer. Many foods go bad very quickly without preservatives. Wastage of food between harvesting and eating is still a problem in many countries.	E211 – sodium benzoate
E3 . . .	antioxidants	Help to stop food reacting with oxygen. Oxygen in the air affects many foods badly, making it impossible to eat them. A good example of this is what happens when you cut an apple open – the brown colour formed is due to oxygen reacting with the apple.	E300 – vitamin C
E4 . . .	emulsifiers, stabilisers and thickeners	Help to improve the texture of the food – what it feels like in your mouth. Many foods need to be treated like this, for example, jam and the soya proteins used in veggieburgers.	E440 – pectin
E5 . . .	acidity regulators	Help to control pH. The acidity of foods is an essential part of their taste. All fruits contain sugar, but without acids they would be sickly and dull.	E501 – potassium carbonate
E6 . . .	flavourings	There are really only five flavours – *sweet*, *sour*, *bitter*, *salt* and *savoury*. What we call flavour is a subtle blend of these five, together with the smells that foods give off.	E621 – monosodium glutamate

Detecting additives

There is a wide range of chemical instruments that scientists can use to identify unknown chemical compounds, including food additives. Many of these are simply more sophisticated and automated versions of techniques that we use in the school lab.

One good example of a technique used to identify food additives is **chromatography**. This technique separates different compounds based on how well they dissolve in a particular solvent. Their solubility then determines how far they travel across a surface, like a piece of chromatography paper.

PRACTICAL

Detecting dyes in food colourings

Make a chromatogram to analyse various food colourings.

- What can you deduce from your chromatogram?

Figure 2 The technique of paper chromatography that we use at school. Although they are more complex, techniques used to identify food additives are often based on the same principles as the simple tests we do in the school science lab.

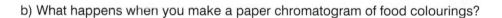

b) What happens when you make a paper chromatogram of food colourings?

Once the compounds in a food have been separated out using chromatography, they can be identified by comparing them with known substances. Alternatively they may be fed into another instrument – the **mass spectrometer**. This can be used for identifying both elements and compounds – it measures the relative formula mass of substances placed in it for analysis, which we can then use to identify the sample.

Figure 3 In the UK in 2005 a batch of red food colouring was found to be contaminated with a chemical suspected of causing cancer. This dye had found its way into hundreds of processed foods. All of these had to be removed from the shelves of our supermarkets and destroyed.

SUMMARY QUESTIONS

1 Copy and complete the table:

Additive	Reason	Additive	Reason
colouring	...	...	improving texture
...	help food keep longer	acidity regulators	...
antioxidants	...	...	changing flavour

2 a) Carry out a survey of some processed foods. Identify some examples of food additives and explain why they have been used.
   b) Describe how we can separate the dyes in a food colouring and identify them.

KEY POINTS

1 Additives may be added to food in order to improve its appearance, taste and how long it will keep (its shelf-life).
2 Food scientists can analyse foods to identify additives.

C1b 5.5 Vegetable oils as fuels

LEARNING OBJECTIVES

1 How can we use vegetable oils as a fuel?
2 What are the advantages of using vegetable oils as fuels?

Fish 'n' chip fuel

Back Forward Stop Refresh Home Autofill Print Mail

Address:

NEWS

Fish 'n' chip shop on wheels

A man in Wales has converted his diesel car to run on old oil from fish and chip shops. Brian Sadler reckons that this could cut the cost of running his car. He regularly visits his local chippies to stock up on fuel supplies. 'Of course, you need to filter it to take the old bits of chips out!' he says.

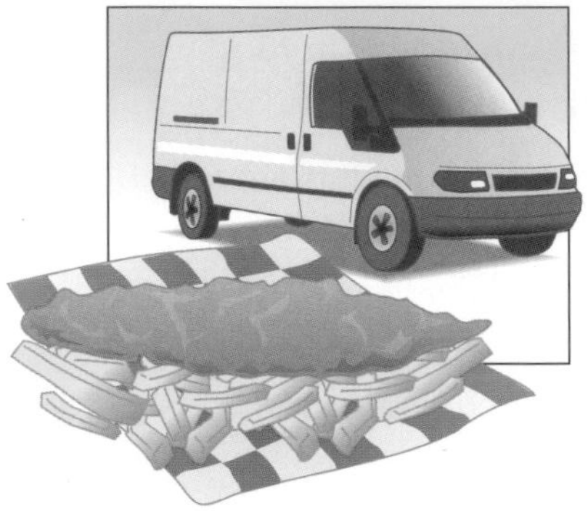

Figure 1 A true story of recycling

Running our cars on the old oil from fish and chip shops isn't realistic for most of us. But it is possible to make fuel for cars using vegetable oils, even used cooking oil. We just need to use a little bit of clever chemistry.

Biodiesel is the name we give to any fuel made from vegetable oils. We can use these fuels in any car or van that has a diesel engine.

Most modern biodiesel is made by treating vegetable oils to remove some unwanted chemicals. The biodiesel made like this can be used on its own, or mixed with diesel fuel made from refining crude oil.

When we make biodiesel we also produce other useful products. For example, we get a solid waste material that we can feed to cattle as a high-energy food. We also get glycerine which we can use to make soap.

a) What is biodiesel?
b) What is biodiesel made from?

There are some very big advantages in using biodiesel as a fuel. First, biodiesel is a very clean fuel. Biodiesel is much less harmful to animals and plants than diesel made from crude oil. If it is spilled, it breaks down about five times faster than 'normal' diesel. Also, when we burn biodiesel in an engine it burns much more cleanly. It makes very little sulfur dioxide and other pollutants.

DID YOU KNOW...

. . . that the largest factory for turning used cooking oil into biodiesel was opened in Scotland in 2005? The plant can produce up to 5% of Scotland's diesel fuels requirements.

But using biodiesel has one really big advantage over petrol and diesel. It is the fact that the crops used to make biodiesel absorb carbon dioxide gas as they grow. So biodiesel is effectively 'CO_2 neutral'. That means the amount of carbon dioxide given off is nearly balanced by the amount absorbed. Therefore, biodiesel makes little contribution to the greenhouse gases in our atmosphere.

Figure 2 This coach runs on biodiesel

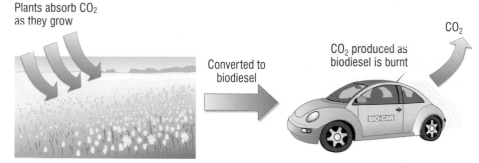

Figure 3 Cars run on biodiesel produce very little CO_2 overall, as CO_2 is absorbed by plants as the fuel is made

Using ethanol as a bio-fuel

Another bio-fuel is ethanol. We can make it by fermenting the sugar from sugar beet or sugar cane. In Brazil they can grow lots of sugar cane. They add the ethanol to petrol, saving our dwindling supplies of crude oil. The ethanol gives off carbon dioxide (a greenhouse gas) when it burns, but the sugar cane absorbs the gas during photosynthesis.

Ethanol can also be made from the ethene we get by cracking the heavier fractions from crude oil (see page 70):

$$C_2H_4 + H_2O \xrightarrow[\text{high pressure}]{\text{catalyst}} C_2H_5OH$$

ethene   steam   ethanol

c) Why is ethanol from sugar cane known as a 'bio-fuel' whereas ethanol from ethene isn't?

Figure 4 Ethanol can be made from sugar cane

SUMMARY QUESTIONS

1 Copy and complete using the words below:

carbon dioxide   cattle   diesel   plants   soap

Biodiesel is a fuel made from .... Making the fuel also produces some other useful products, including ...... and food for .... Biodiesel produces less pollution than ...... , and absorbs nearly as much .... when it is made as it does when it burns.

2 Where does the energy in biodiesel come from?

3 a) How is ethene converted into ethanol? Include a balanced equation in your answer.
b) When can we describe ethanol as a bio-fuel?
c) Write an article for a local newspaper describing the arguments for using biodiesel instead of other fuels made from crude oil.

KEY POINTS

1 Vegetable oils can be burned as fuels.
2 Vegetable oils are a renewable source of energy that could be used to replace some fossil fuels.

C1b 5.6   Vegetable oils

Plant-powered performance

Whatever will those ace scientists come up with next? Just when you think you've seen everything, the idea of filling up your car with plant power comes along! The latest idea by those who want us all to save the planet is to take oil made from oilseed rape plants (you know, that yellow stuff you see growing all over the countryside in springtime), mix it with diesel and then shove it in the tank of your motor! And they call it ... *biodiesel!*

So why would anyone want to do this? Well, mainly, because biodiesel is better for the environment. Not only does it burn just like ordinary diesel fuel, but it doesn't produce as much pollution. And what's more, it's a 'green fuel' which doesn't contribute as much to global warming – though we couldn't quite see why ... there's a year's free subscription on offer to anyone who can explain it to us!

ACTIVITY

Imagine that you are a presenter on a popular television programme about cars and driving. Write and present an article for this programme about biodiesel.

Make your article fun and informative. Include as much factual information as you can so that people can decide whether they would like to use this fuel in their car.

Look after your family's hearts

Everyone knows the benefits of a healthy diet. But do you know the benefits of ensuring that you eat vegetable oils as part of your diet?

Scientists have found that eating vegetable oils instead of animal fats can do wonders for the health of your heart. The saturated fats you find in things like butter and cheese can make the blood vessels of your heart become clogged up with fat.

However, the unsaturated fats in vegetable oils (like olive oil and corn oil) are especially good for you. They help to keep your arteries clear and reduce the chance of you having heart disease.

The levels of a special fat called **cholesterol** in your blood give doctors an idea about your risk of heart disease. People who eat lots of vegetable oils tend to have a much healthier level of cholesterol in their blood.

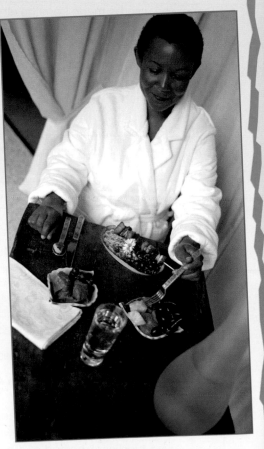

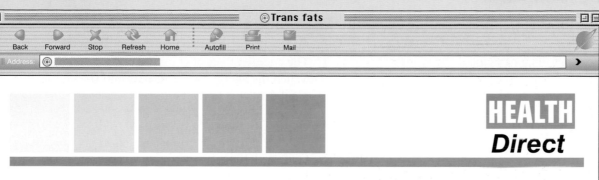

HEALTH
Direct

Online Health Encyclopaedia
Trans fats

Trans unsaturated fatty acids, or ***trans fats***, are solid fats. They are produced artificially by heating liquid vegetable oils with metal catalysts and hydrogen. Trans fats are made in huge quantities when we harden vegetable oils to make margarine.

The fats used to cook French fries and other fast foods usually contain trans fats. Concerns have been raised for several decades that eating trans fats might have contributed to the 20th century epidemic of coronary heart disease.

Studies have shown that trans fats have adverse effects on cholesterol in the blood – increasing 'bad' cholesterol while decreasing 'good' cholesterol. Trans fats have also been associated with an increased risk of coronary heart disease in other studies.

Changes in food labelling are very important. But many products, including fast food, often contain extremely high levels of trans fats. Yet these are exempt from labelling regulations and may have labels such as 'cholesterol-free' and 'cooked in vegetable oil'.

For example, a person eating one doughnut for breakfast and a large order of French fries for lunch would eat 10 g of trans fats, or 5 per cent of the total energy of an 1800-calorie diet. Thus, simple labelling changes alone will not be sufficient.

ACTIVITIES

a) Write an article for a family lifestyle magazine about 'feeding your family'. Include in this article reasons for including vegetable oils in a balanced diet and their effect on people's health.

b) Design a poster with the title 'Vegetable oils – good or bad?'

SUMMARY QUESTIONS

1 Write simple definitions for the following words:

   a) vegetable oils

   b) unsaturated oils

   c) saturated oils.

2 A vegetable oil removes the colour from bromine solution. When the oil has been hardened it does not react with bromine solution.
   Explain these observations.

3 a) Some ice cream is left standing out on a table during a meal on a hot day. It is then put back in the freezer again. When it is taken out of the freezer a few days later, people complain that the ice cream tastes 'crunchy'. Why is this?

   b) A recipe for ice cream says: 'Stir the ice cream from time to time while it is freezing.' Why must you stir ice cream when freezing it?

4 a) Why is it important to be able to identify food additives in food?

   b) What is the first number of the E number of the following additives?

   i) lecithin (an emulsifier)

   ii) quinoline yellow (a food colouring)

   iii) calcium oxide (an acidity regulator)

   iv) inosinic acid (a flavour enhancer)

5 Biodiesel is almost 'carbon neutral'. Explain, using your knowledge of the carbon cycle, whether it would ever be possible to produce biodiesel that produces less carbon dioxide when burnt than was absorbed by the plants which made it.

6 'Alternative fuels like biodiesel and gasohol are all very well for countries like Brazil and India – but no good in countries like the UK!' Do you agree with this statement? Give your reasons.

7 Some food are marketed as 'free from artificial additives'.

   a) Does this mean that the food contains no added substances?

   b) Give an example of a substance that is added to food, and outline the reasons for using it.

   c) Is 'additive-free food' better than food which contains additives?

EXAM-STYLE QUESTIONS

1 The energy values of chips depend on their fat content. Match the energy values **A**, **B**, **C** and **D** with the numbers **1** to **4** in the table.

   A 687 kJ/100 g

   B 796 kJ/100 g

   C 1001 kJ/100 g

   D 1174 kJ/100 g

	Description of type of chips	Fat content (g/100 g)
1	Fish and chip shop, fried in blended oil	12.4
2	French fries from burger outlet	15.5
3	Homemade fried in blended oil	6.7
4	Oven chips, frozen, baked	4.2

(4)

2 Match the words **A**, **B**, **C** and **D** with spaces **1** to **4** in the sentences.

   A cooking oils

   B emulsifiers

   C emulsions

   D hydrogenated oils

   Mayonnaise and salad dressings are ......**1**...... that are made by mixing oil and vinegar with other ingredients such as egg yolk.

   In mayonnaise the egg yolk contains ......**2**......that stop the oil and water separating.

   Vegetable oils can be converted into ......**3**...... by reacting with hydrogen and a catalyst.

   Biodiesel is a fuel that can be made from waste ......**4**...... . (4)

3 The table on the next page gives some information about four different vegetable oils.
   Smoke point is the temperature at which the oil begins to smoke when heated.
   Match descriptions **A**, **B**, **C** and **D** with numbers **1** to **4** in the table.

   A The oil that contains the most monounsaturated fat.

   B The oil that reacts with the largest volume of bromine water.

   C The oil with the highest melting point.

   D The oil with the widest range of smoke point.

	Type of oil			
	1	**2**	**3**	**4**
	Corn oil	**Olive oil**	**Sunflower oil**	**Rapeseed oil**
Saturated fat (%)	14.4	14.3	12.0	6.6
Mono-unsaturated fat (%)	29.9	73.0	20.5	59.3
Poly-unsaturated fat (%)	51.3	8.2	63.3	29.3
Melting point (°C)	−15	−12	−18	5
Smoke point (°C)	229–268	204–210	229–252	230–240

(4)

4 Use the table of types of chips in question **1** to help you answer these questions.

(a) Why do chips contain fat? (1)

(b) Why do French fries contain most fat? (1)

(c) Why do oven chips contain least fat? (2)

(d) Why do all chips have a golden brown colour, but boiled potatoes remain white? (2)

(e) (i) How would you display the data in the table? (1)
    (ii) Explain your answer to part (i). (1)

5 Virgin olive oil is extracted by mechanical methods that do not modify its properties. If the temperature during extraction does not exceed 27°C the oil can be labelled as 'cold pressed'. Any olive oil that remains in the pressed pulp can be extracted by dissolving it in a solvent. The solvent is removed from the oil by evaporation. This type of oil is called pomace oil.

(a) Why is it important that the temperature does not exceed 27°C during extraction? (2)

(b) Suggest why some people prefer virgin olive oil to pomace oil. (2)

6 Some students made a solution of the colours in a soft drink. Describe how you could use paper chromatography to show how many colours were in the solution. (4)

HOW SCIENCE WORKS QUESTIONS

The teacher decided that her class should do a survey of different foods to find out the degree of unsaturated oils present in them. She chose five different oils and divided them amongst her students. This allowed one oil to be done twice, by two different groups. They were given strict instructions as to how to do the testing.

Bromine water was added to each oil from a burette. The volume added before the bromine was no longer colourless was noted.

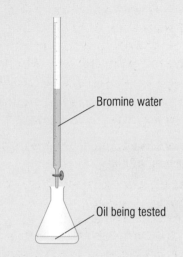

Bromine water

Oil being tested

The results are in this table.

Type of oil	Amount of bromine water added (cm³)	
	Group 1	**Group 2**
Ollio	24.2	23.9
Soleo	17.8	18.0
Spreo	7.9	8.1
Torneo	13.0	12.9
Margeo	17.9	17.4

a) Why was it important that the teacher gave strict instructions to all of the groups on how to carry out the tests? (1)

b) List some controls that should have been included in the instructions. (3)

c) Are there any anomalies? If so, state which results are anomalies. (1)

d) What evidence is there in the results that indicate that they are reliable? (1)

e) How might the accuracy be checked? (1)

f) How would you present these results? (1)

C1b 6.1   Structure of the Earth

LEARNING OBJECTIVES

1 What is below the surface of the Earth?
2 How did people used to think that mountains and valleys were formed?

The deepest mines go down to about 3500 m, while geologists have drilled down to more than 12 000 m in Russia. Although these figures seem large, they are tiny compared with the diameter of the Earth. The Earth's diameter is about 12 800 km. That's more than one thousand times the deepest hole ever drilled!

The Earth is made up of layers that formed many millions of years ago, early in the history of our planet. Heavy matter sank to the centre of the Earth while lighter material floated on top. This produced a structure consisting of a dense **core**, surrounded by the **mantle**. Outside the mantle there is a thin layer called the **crust**.

The uppermost part of the mantle and the crust make up the Earth's *lithosphere*.

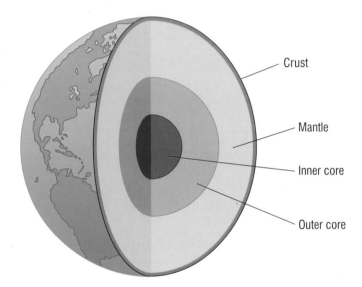

Figure 1 The structure of the Earth

We call the outer layer of the Earth the crust. This layer is very thin compared to the diameter of the Earth (thinner than the outer layer of a football!). Its thickness can vary from as thin as 5 km under the oceans to as much as 70 km under the continents.

Underneath the crust is the mantle. This layer is much, much thicker than the crust – nearly 3000 km. The mantle behaves like a solid, but it can flow in parts very slowly.

Finally, inside the mantle lies the Earth's core. This is about half the radius of the Earth, and is made of a mixture of nickel and iron. The core is actually two layers. The outer core is a liquid, while the inner core is solid.

a) What is the outer layer of the Earth called?
b) What is the next layer of the Earth called?
c) How many parts make up the Earth's core?

How do we know the structure inside the Earth if we have never seen it? Scientists use evidence from earthquakes. Following an earthquake, seismic waves travel through the Earth. The way in which seismic waves travel through the Earth is affected by the structure of the Earth. By observing how seismic waves travel, scientists have built up the detailed picture of the inside of the Earth described here.

Also, by making careful measurements, physicists have been able to measure the mass of the Earth, and to calculate its density. The density of the Earth as a whole is much greater than the density of the rocks found in the crust. This suggests that the centre of the Earth must be made from a different material to the crust. This material must have a much greater density than the material that makes up the crust.

Crust	Mantle	Core
Averages: about 6 km under the oceans about 35 km under continental areas	Starts underneath crust and continues to about 3000 km below Earth's surface Behaves like a solid, but is able to flow very slowly	Radius of about 3500 km Made of nickel and iron Outer core is liquid, inner core is solid

At one time scientists thought that features like mountain ranges on the surface of the Earth were caused by the crust shrinking as the early molten Earth cooled down. They thought of it rather like the skin on the surface of a bowl of custard shrinks then wrinkles as the custard cools down.

However, scientists now have a better explanation for the features on the Earth's surface, as we shall see later in this chapter.

Figure 2 All of the minerals that we depend on in our lives – iron, aluminium and copper, for example, as well as oil and gas – come from the thin crust of the Earth

SUMMARY QUESTIONS

1 Copy and complete using the words below:

   core   crust   mantle   slowly   solid   thin

   The structure of the Earth consists of three layers – the ……, the …… and the …… . The outer layer of the Earth is very …… compared to its diameter. The layer below this is ……. but can flow in parts very …….

2 Why do some people think that the mantle is best described as a 'very thick syrupy liquid'?

3 Why do scientists think that the core of the Earth is made of much denser material than the crust?

KEY POINTS

1 The Earth consists of a series of layers.
2 Scientists originally thought that the features on the Earth's surface were caused as the crust cooled and shrank.

C1b 6.2 The restless Earth

LEARNING OBJECTIVES

1 What are tectonic plates?
2 Why do they move?
3 Why is it difficult for scientists to predict when earthquakes and volcanic eruptions will occur?

Figure 1 *Glossopteris* was a tree-like plant growing about 230 million years ago. It had tongue-shaped leaves, and grew to a height of about 4 metres. Its fossils have been found in Africa and in South America.

Have you ever looked at the western coastline of Africa and the eastern coastline of South America on a map? If you have, you might have noticed that these edges of the two continents have a remarkably similar shape.

The fossils and rock structures that we find when we look in Africa and South America are also similar. Fossils show that the same reptiles and plants once lived in both continents. And the layers of rock in the two continents are arranged in the same sequence, with layers of sandstone lying above seams of coal.

Scientists now believe that they can explain the similarity in the shapes of the continents and of the rocks and fossils found there. They think that the two continents were once joined together as one land mass.

a) What evidence is there that Africa and South America were once joined?

Figure 2 shows the vast 'supercontinent' of Pangaea. This land mass is believed to have existed up until about 250 million years ago. Slowly Pangaea split in two about 160 million years ago. The land masses continued to move apart until about 50 million years ago. Then they began to closely resemble the map of the world we know today.

b) What was the name of the original 'supercontinent'?

Of course, the continents moved and split up very, very slowly – only a few centimetres each year. They moved because the Earth's crust and uppermost part of the mantle (its lithosphere) is cracked into a number of large pieces. We call these **tectonic plates**.

Deep within the Earth, radioactive decay produces vast amounts of energy. This heats up molten minerals in the mantle which expand. They become less dense and rise towards the surface and are replaced by cooler material. It is these **convection currents** which pushed the tectonic plates over the surface of the Earth.

Figure 2 The break up of Pangaea into Laurasia and Gondwanaland led eventually to the formation of the land masses we recognise today. Notice how, 100 million years ago, India is still moving northwards to take up the position it occupies today. The collision between India and the continent of Asia produced the mountain range we call the Himalayas.

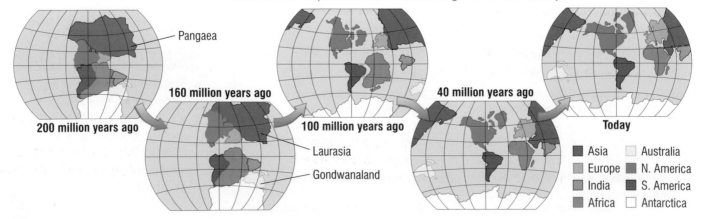

Where the boundaries of the plates meet, huge forces are exerted. These forces make the plates buckle and deform, and mountains may be formed. The plates may also move suddenly and very quickly past each other. These sudden movements cause earthquakes.

If earthquakes happen under the sea, they may cause huge tidal waves called **tsunamis**. However, it is difficult for scientists to know exactly where and when the plates will move like this. So predicting earthquakes is still a very difficult job.

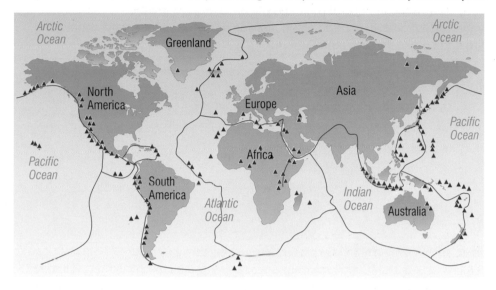

Figure 3 The distribution of volcanoes around the world largely follows the boundaries of the tectonic plates

GET IT RIGHT!

The Earth's tectonic plates are made up of the crust and the upper part of the mantle (not just the crust).

Wegener's revolutionary theory

The idea that huge land masses once existed before the continents we know today was put forward in the late 19th century by the geologist Edward Suess. He thought that a huge southern continent had sunk. He suggested that this left behind a land bridge (since vanished) between Africa and South America.

The idea of continental drift was put first forward by Alfred Wegener in 1915. However, his fellow scientists found Wegener's ideas hard to accept. This was mainly because he could not explain *how* the continents had moved. So they stuck with their existing ideas.

His theory was finally shown to be right almost 50 years later. Scientists found that the sea floor is spreading apart in some places, where molten rock is spewing out between two continents. This led to a new theory, called plate tectonics.

SUMMARY QUESTIONS

1 Copy and complete using the words below:

   convection   earthquakes   mantle   tectonic   tsunamis   volcanoes

   The surface of the Earth is split up into a series of ...... plates. These move across the Earth's surface due to ...... currents in the .... Where the plates meet or rub against each other ...... and ...... may form, and ...... may happen.

2 a) Explain how tectonic plates move.
   b) Why are earthquakes and volcanic eruptions difficult to predict?
   c) Imagine that you are a scientist who has just heard Wegener talking about his ideas for the first time. Write a letter to another scientist explaining what Wegener has said and why you have chosen to reject his ideas.

KEY POINTS

1 The Earth's lithosphere is cracked into a number of pieces (tectonic plates) which are constantly moving.
2 The motion of the tectonic plates is caused by convection currents in the mantle, due to radioactive decay.
3 Earthquakes and volcanoes happen where tectonic plates meet. It is difficult to know when the plates may slip past each other. This makes it difficult to predict accurately when and where earthquakes will happen.

C1b 6.3

The Earth's atmosphere in the past

Scientists think that the Earth was formed about 4.5 billion years ago. To begin with it was a molten ball of rock and minerals. For its first billion years it was a very violent place. The Earth's surface was covered with volcanoes belching fire and smoke into the atmosphere.

Figure 1 Volcanoes moved chemicals from inside the Earth to the surface and the newly forming atmosphere

The volcanoes released carbon dioxide, water vapour and nitrogen gas, which formed the early atmosphere. Water vapour in the atmosphere condensed as the Earth gradually cooled down, and fell as rain. This collected to form the first oceans.

Comets also brought water to the Earth. As icy comets rained down on the surface of the Earth they melted, adding to the water supplies. Even today many thousands of tonnes of water fall onto the surface of the Earth from space every year.

So as the Earth began to stabilise, the early atmosphere was probably mainly carbon dioxide, with some water vapour, nitrogen and traces of methane and ammonia. There was very little or no oxygen. This is very like the atmospheres which we know exist today on the planets Mars and Venus.

a) What was the main gas in the Earth's early atmosphere?
b) How much oxygen was there in the Earth's early atmosphere?

Figure 2 The surface of one of Jupiter's moons called Io, with its small atmosphere and active volcanoes. This photograph probably gives us a reasonable glimpse of what our own Earth was like billions of years ago.

After the initial violent years of the history of the Earth, the atmosphere remained relatively stable. That's until life first appeared on Earth.

Scientists think that life on Earth began about 3.4 billion years ago, when simple organisms like bacteria appeared. These could make food for themselves, using the breakdown of other chemicals as a source of energy.

Later, bacteria and other simple organisms such as algae evolved. They could use the energy of the Sun to make their own food by photosynthesis, and oxygen was produced as a waste product.

By two billion years ago the levels of oxygen were rising steadily as algae and bacteria filled the seas. More and more plants evolved, all of them also photosynthesising, removing carbon dioxide and making oxygen.

$$\text{carbon dioxide} + \text{water} \xrightarrow{\text{(energy from sunlight)}} \text{sugar} + \text{oxygen}$$

As plants evolved and successfully colonised most of the surface of the Earth, the atmosphere became richer and richer in oxygen. Now it was possible for animals to evolve. These animals could not make their own food and needed oxygen to respire.

On the other hand, many of the earliest living microorganisms could not tolerate oxygen (because they had evolved without it). They largely died out, as there were fewer and fewer places where they could live.

Figure 3 Some of the first photosynthesising bacteria probably lived in colonies like these stromatolites. They grew in water and released oxygen into the early atmosphere.

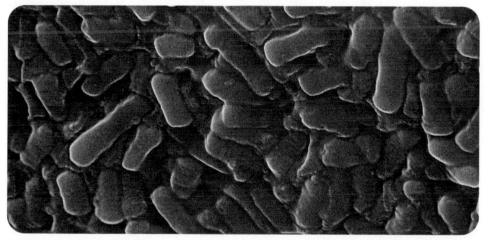

Figure 4 Bacteria such as these not only do not need oxygen – they die if they are exposed to it. But they can survive and breed in rotting tissue and other places where there is no oxygen.

SUMMARY QUESTIONS

1 Copy and complete using the words below:

   carbon dioxide   methane   oxygen   volcanoes   water

   The Earth's early atmosphere probably consisted mainly of the gas .... .There could also have been ...... vapour and nitrogen, plus small amounts of ...... and ammonia. These gases were released by ...... as they erupted. Plants removed carbon dioxide from the atmosphere and produced ....

2 How was the Earth's early atmosphere formed?

3 Why was there no life on Earth for several billion years?

4 Draw a chart that explains the early development of the Earth's atmosphere.

KEY POINTS

1 The Earth's early atmosphere was formed by volcanic activity.

2 It probably consisted mainly of carbon dioxide. There may also have been water vapour together with traces of methane and ammonia.

3 As plants colonised the Earth, the levels of oxygen in the atmosphere rose.

C1b 6.4   Gases in the atmosphere

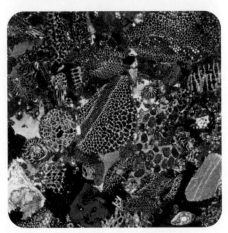

Figure 1 There is clear evidence in carbonate rocks of the organisms which lived millions of years ago, now preserved with their ancient carbon in the structure of our rocks

We think that the early atmosphere of the Earth contained a great deal of carbon dioxide. Yet the modern atmosphere of the Earth has only around 0.04% of this gas. Where has it all gone? The answer is mostly into living organisms and into materials formed from living organisms.

Carbon dioxide is taken up by plants during photosynthesis and the carbon can end up in new plant material. Then animals eat the plants and the carbon is transferred to the animal tissues, including bones, teeth and shells.

Over millions of years the dead bodies of huge numbers of these living organisms built up at the bottom of vast oceans. Eventually they formed sedimentary carbonate rocks like limestone.

Some of these living things were crushed by movements of the Earth and heated within the crust. They formed fossil fuels such as coal and oil. In this way much of the carbon from carbon dioxide in the ancient atmosphere became locked up within the Earth's crust.

a) Where has most of the carbon dioxide in the Earth's early atmosphere gone?

Carbon dioxide also dissolved in the oceans. It reacted and made insoluble carbonate compounds. These fell to the sea-bed and helped to form carbonate rocks.

At the same time, the ammonia and methane, from the Earth's early atmosphere, reacted with the oxygen formed by the plants. This got rid of these poisonous gases and increased the nitrogen and carbon dioxide levels:

$$CH_4 + 2O_2 \rightarrow CO_2 + 2H_2O$$
$$4NH_3 + 3O_2 \rightarrow 2N_2 + 6H_2O$$

By 200 million years ago the proportions of the different gases in the Earth's atmosphere were much the same as they are today.

Look at the pie chart in Figure 2.

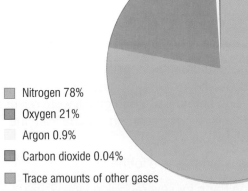

- Nitrogen 78%
- Oxygen 21%
- Argon 0.9%
- Carbon dioxide 0.04%
- Trace amounts of other gases

Figure 2 The relative proportions of nitrogen, oxygen and other gases in the Earth's atmosphere

b) What gas did plants produce that changed the Earth's atmosphere?

The noble gases

The Earth's atmosphere contains tiny amounts of a group of gases that we call the **noble gases**. They are all found in Group 0 of the periodic table.

Helium, neon and argon, along with krypton, xenon and radon are the least reactive elements known. It is very difficult to make them react with any other elements. They don't even react with themselves to form molecules. They exist as single atoms. We say that they are monatomic.

Because the noble gases are very unreactive, we cannot use them to make useful materials. Instead, we use them in situations where they are useful because of their extreme lack of reactivity.

Figure 3 The noble gases are all found in Group 0 of the periodic table

Uses of the noble gases

Helium is used in airships and in party balloons. Its low density means that balloons filled with the gas float in air. It is also safer than the only alternative gas, hydrogen, because its low reactivity means that it does not catch fire. We also use it with oxygen as a breathing mixture for deep-sea divers. The mixture reduces their chances of suffering from the 'bends'.

We use **neon** in electrical discharge tubes – better known as **neon lights**. When we pass an electrical current through the neon gas it gives out a bright light. Neon lights are familiar as street lighting and in advertising.

We use **argon** in a different type of lighting – the everyday light bulb (or filament lamp). The argon provides an inert atmosphere inside the bulb. When the electric current passes through the metal filament, the metal becomes white hot. If any oxygen was inside the bulb, it would react with the hot metal. However, no chemical reaction takes place between the metal filament and argon gas. This stops the filament from burning away and makes light bulbs last longer.

Figure 4 These brightly coloured balloons are filled with helium, which makes them float upwards through the air

---

SUMMARY QUESTIONS

1 Copy and complete the table showing the proportion of gases in the Earth's atmosphere today.

nitrogen	oxygen	argon	carbon dioxide	other gases
%	%	%	%	%

2 How did the evolution of plants change the Earth's atmosphere?

3 a) Explain how ammonia and methane were probably removed from the Earth's atmosphere.
   b) Find out how the noble gases are used and make a list that explains at least one use of each gas.

KEY POINTS

1 The main gases in the Earth's atmosphere are oxygen and nitrogen.
2 About four-fifths (80%) of the atmosphere is nitrogen, and one-fifth (20%) is oxygen.
3 The noble gases are unreactive gases found in Group 0 of the periodic table. Their lack of reactivity makes them useful in many ways.

C1b 6.5   The carbon cycle

Over the past 200 million years the levels of carbon dioxide in the atmosphere have not changed much. This is due to the natural carbon cycle in which carbon moves between the oceans, rocks and the atmosphere.

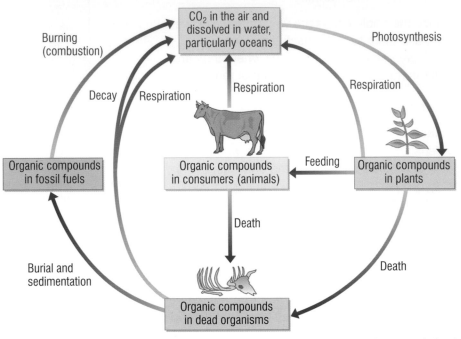

Figure 1 The carbon cycle has kept the level of carbon dioxide in the atmosphere steady for the last 200 million years

Left to itself, the carbon cycle is self-regulating. The oceans act as massive reservoirs of carbon dioxide. They absorb excess CO_2 when it is produced and release it when it is in short supply. Plants also soak up carbon dioxide from the atmosphere. We often call plants and oceans carbon dioxide 'sinks'.

Carbon dioxide moves back into the atmosphere when living things respire or when they die and decompose.

Then we also have the CO_2 that comes from volcanoes. Carbonate rocks are sometimes moved deep into the Earth by the movements of the Earth's crust. If that rock then becomes part of a volcano, heat causes the carbonates in the rock to break down. Then the carbon dioxide gas is released as the volcano erupts.

a) What has kept carbon levels stable over the last 200 million years?
b) What happens when carbonate rocks become part of a volcano?

PRACTICAL

The properties of carbon dioxide

Carry out a series of tests to find out the properties of carbon dioxide gas.

● Record your findings in a bullet-pointed list.

The changing balance

Over the last fifty years or so we have increased the amount of carbon dioxide released into the atmosphere tremendously. We burn fossil fuels to make electricity, heat our homes and drive our cars. This has enormously increased the amount of carbon dioxide we produce.

There is no doubt that the levels of carbon dioxide in the atmosphere are increasing.

We can record annual changes in the levels of carbon dioxide which are due to seasonal differences in the plants. The variations within each year show how important plants are for removing CO_2 from the atmosphere. But the overall trend for the last 30 years has been ever upwards.

The balance between the carbon dioxide produced and the carbon dioxide absorbed by 'CO_2 sinks' is very important.

Think about what happens when we burn fossil fuels. Carbon, which has been locked up for hundreds of millions of years in the bodies of once-living animals, is released as carbon dioxide into the atmosphere. For example:

$$\text{propane} + \text{oxygen} \rightarrow \text{carbon dioxide} + \text{water}$$
$$C_3H_8 + 5O_2 \rightarrow 3CO_2 + 4H_2O$$

As carbon dioxide levels in the atmosphere go up, so the reaction between carbon dioxide and sea water increases. This reaction makes insoluble carbonates (mainly calcium carbonate). These are deposited as sediment on the bottom of the ocean. It also produces soluble hydrogencarbonates – mainly calcium and magnesium – which simply remain dissolved in the sea water.

In this way the seas and oceans act as a buffer, absorbing excess carbon dioxide but releasing it if necessary. However this buffering system probably cannot cope with all the additional carbon dioxide that we are currently pouring out into the atmosphere.

Figure 2 Most of the electricity that we use in the UK is made by burning fossil fuels. This releases carbon dioxide into the atmosphere. One solution would be to pump the CO_2 deep underground to be absorbed into porous rocks. This would increase the cost of producing electricity by about 10%.

> ### DID YOU KNOW...
>
> . . . that scientists predict that global warming may mean that the Earth's temperature could rise by as much as 5.8°C by the year 2100?

SUMMARY QUESTIONS

1 Match up the parts of sentences:

a) Carbon dioxide levels in the Earth's atmosphere	A ...... carbon locked up long ago is released as carbon dioxide.
b) Plants and oceans are known as	B ...... were kept steady by the carbon cycle.
c) When we burn fossil fuels	C ...... the reaction between carbon dioxide and sea water increases.
d) As carbon dioxide levels rise	D ...... carbon dioxide sinks

2 Draw a labelled diagram to illustrate how boiling an electric kettle may increase the amount of carbon dioxide in the Earth's atmosphere.

3 Why has the amount of carbon dioxide in the Earth's atmosphere risen in the last 50 years?

> ### KEY POINTS
>
> 1 Carbon moves into and out of the atmosphere due to plants, the oceans and rocks.
> 2 The amount of carbon dioxide in the Earth's atmosphere has risen due to the amount of fossil fuels we burn.

EARTHQUAKES

The tsunami that tore across the Indian Ocean on 26 December 2004 left nearly 300 000 people dead. It was caused when a huge earthquake – the second biggest ever recorded – lifted billions of tonnes of sea water over 20 metres upwards. This produced a huge wave that travelled thousands of miles at speeds of several hundred miles an hour.

The earthquake was detected thousands of miles away by the Pacific Tsunami Warning Centre, but no-one could tell that it would produce such devastation. In the aftermath of the disaster, people discussed the lessons that had been learnt.

DEATH TOLL FROM TSUNAMI MAY REACH 300,000
six page report

MASSIVE EARTHQUAKE CAUSES TIDAL WAVE IN INDIAN OCEAN
Full report

What we need to do is to build an early warning system that would tell us when a tidal wave is coming. A system like this already exists in the Pacific Ocean – but now we need one in the Indian Ocean.

Rather than detect tidal waves it would be better to predict when an earthquake was going to happen – then we could help to protect people against earthquakes and tidal waves.

Any early warning system would cost millions of dollars. Tidal waves don't happen that often. The money would be better spent on feeding people who are starving and buying medicines to treat people who are sick.

ACTIVITY

Set up a discussion between three people on a TV news show or a radio phone-in. Each person should take one of the viewpoints described above, and should argue for their particular point of view.

Try to find as much factual information as you can to back up your view:

● whether you think a tsunami early warning system is needed, or
● that money is better spent on food and health, or
● that the effort should be put into detecting earthquakes.

THE CARBON PROBLEM

Our lifestyles all affect the Earth – and in particular, the amount of carbon dioxide we produce from the fuels we burn and the electricity we use. Here are four different people – you might even know one or two of them . . .

> It's sooo difficult just keeping up these days. The right clothes, makeup and hair – not to mention being seen at all the right places. Fortunately Mummy has lots of time to take me to where I have to be in her new car ... cool isn't it? Must dash – I hear they've got some new jeans in at that lovely little shop, only 30 miles drive away! Mummy ... !

> It's alright round here I suppose – but not a lot to do except watch telly and play computer games. I'd like to go up to London to see the football at the weekend but I can't afford it and Dad says it's too far for him to drive. S'pose I'll just have to listen to it on the radio ...

> It's amazing what cheap air tickets you can get now ... I'm just off to see a mate in California! Spent time in Spain, Morocco and Eastern Europe so far this year – and if I can get the money together, I might even manage to get to Australia. Brilliant gap year!

> Haven't got much time to stop and talk – just on my way to the 'save the whale' rally. It's 20 miles away, but it won't take me long to get there on my trusty bike. Like my new jeans – cool label, but from the charity shop! Wanna share a tofu sandwich ... !

ACTIVITY

Look at each of these characters. Choose **one** of them and write a 'diary' for what they do and where they go in a typical week. When you have done this, think about the amount of fossil fuels this character uses, not just in travel but in everything they do and buy. Make a list of ways that they could reduce their fossil fuel consumption – and so reduce the amount of carbon dioxide they produce.

SUMMARY QUESTIONS

1 Write simple definitions for the following words describing the structure of the Earth:

a) mantle

b) core

c) lithosphere

d) tectonic plate.

2 Write down *three* pieces of evidence that suggest that South America and Africa were once joined together.

3 The pie charts show the atmosphere of a planet shortly after it was formed (A) and then millions of years later (B).

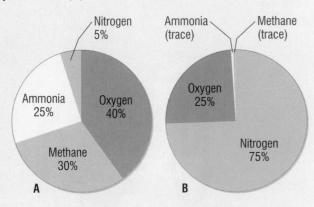

a) How did the atmosphere of the planet change?

b) What might have caused this change?

c) Copy and complete the word equations showing the chemical reactions that may have taken place in the atmosphere.

i) methane + ... → carbon dioxide + ...

ii) ammonia + ... → nitrogen + ...

4 Wegener suggested that all the Earth's continents were once joined in a single land mass.

a) Describe the evidence for this idea, and explain how the single land mass separated into the continents we see today.

b) Why were other scientists slow to accept Wegener's ideas?

5 The Earth and its atmosphere are constantly changing. Design a poster to show this, that would be suitable for displaying in a classroom with children aged 10–11 years.
Use diagrams and words to describe and explain ideas and to communicate them clearly to the children.

EXAM-STYLE QUESTIONS

1 Match words **A**, **B**, **C** and **D** with the numbers **1** to **4** in the table.

A atmosphere

B core

C crust

D mantle

	Description
1	Almost entirely solid, but can flow very slowly
2	Contains mainly the elements nitrogen and oxygen
3	Has an average thickness of about 6 km under oceans and 35 km under continents
4	Part liquid and part solid, with a radius of about 3 500 km

(4)

2 Match words **A**, **B**, **C** and **D** with the spaces **1** to **4** in the sentences.

A believed

B dismissed

C produced

D published

In 1912 Alfred Wegener ......**1**...... a theory that a single land mass had split apart into continents that moved to their current positions.

At the time geologists ......**2**...... that the continents moved up and down – not sideways.

Wegener's theory was ......**3**...... by geologists because he could not explain how the continents moved.

In 1944 an English geologist explained that heat from radioactivity ......**4**...... convection currents strong enough to move continents.　　　　　(4)

3 Match words **A**, **B**, **C** and **D** with spaces **1** to **4** in the sentences.

A ammonia

B carbon dioxide

C noble gases

D oxygen

The Earth's early atmosphere consisted mainly of ......**1**...... with some nitrogen, water vapour, methane and ......**2**...... .

The Earth's atmosphere now contains 78% nitrogen, 21% ......**3**......, about 1% ......**4**....... and 0.04% carbon dioxide.　　　　　(4)

HOW SCIENCE WORKS QUESTIONS

4 This question is about three of the noble gases, helium, neon and argon.

(a) Why is helium used in balloons and airships rather than hydrogen? (2)

(b) Explain how argon allows you to use an electric light bulb for many hours. (2)

(c) Explain how neon is used for advertising. (2)

5 The graph shows the percentage of carbon dioxide in the atmosphere in recent years.

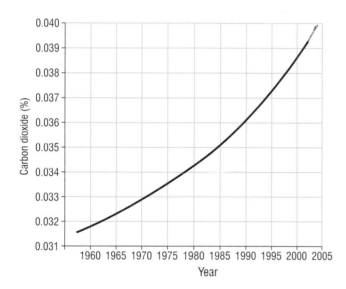

(a) (i) Could the 'percentage of carbon dioxide in the atmosphere' be described as a categoric, discrete or continuous variable? (1)

(ii) There is considerable variation in the percentage of carbon dioxide within each 5 year period. What do we call the line that 'smooths out' these variations on the graph? (1)

(b) By how much has the percentage of carbon dioxide increased from 1960 to 2005? (1)

(c) What is this increase as a percentage of the 1960 figure? (1)

(d) Suggest **two** reasons for this increase. (2)

(e) What natural processes remove carbon dioxide from the atmosphere? (3)

(f) Why should we be concerned about the increase in carbon dioxide? (2)

Core samples have been taken of the ice from Antarctica. The deeper the sample the longer it has been there. It is possible to date the ice and to take air samples from it. The air was trapped when the ice was formed. It is possible therefore to test samples of air that have been trapped in the ice for many thousands of years.

This table shows some of these results. The more recent results are from actual air samples taken from a Pacific island.

Year	CO_2 concentration (ppm)	Source
2005	379	Pacific island
1995	360	Pacific island
1985	345	Pacific island
1975	331	Pacific island
1965	320	Antarctica
1955	313	Antarctica
1945	310	Antarctica
1935	309	Antarctica
1925	305	Antarctica
1915	301	Antarctica
1905	297	Antarctica
1895	294	Antarctica
1890	294	Antarctica

a) If you have access to a spreadsheet, enter this data and produce a line graph. (3)

b) Draw a line of best fit. (1)

c) What pattern can you detect? (2)

d) What conclusion can you make? (1)

e) Should the fact that the data came from two different sources affect your conclusion? Explain why. (2)

EXAMINATION-STYLE QUESTIONS

Chemistry A

1 Match the polymers **A**, **B**, **C** and **D** with the numbers **1** to **4** in the table.

See pages 72–5

    A   high density poly(ethene)

    B   low density poly(ethene)

    C   poly(chloroethene)

    D   poly(propene)

	Density (g/cm³)	Monomer
1	0.90	C_3H_6
2	0.92	C_2H_4
3	0.96	C_2H_4
4	1.30	C_2H_3Cl

(4 marks)

2 Alfred Wegener first suggested his theory of continental drift in 1915. He suggested that the continents had once been a single large land mass that had split apart. He showed that fossils and rocks were similar on the parts of America and Africa that fitted together. He produced a lot of evidence to support his theory but other scientists did not accept his ideas for over 50 years.

See pages 98–9

(a) One of the main reasons why other scientists did not accept his ideas was . . .

    A   America and Africa are too far apart.

    B   Fossils are similar all over the Earth.

    C   He could not explain how the continents moved.

    D   He had no evidence that the continents were moving.

(b) Geologists did not like his ideas because . . .

    A   they could not understand his fossil evidence.

    B   they did not know enough about the rocks of Africa and America.

    C   they did not like German scientists.

    D   they would have to change all their own long established ideas.

(c) What new evidence was found in the 1960s?

    A   Land bridges had existed between the continents.

    B   The sea floor was spreading on either side of deep ocean ridges.

    C   The polar ice-caps were shown to be melting.

    D   The Earth had been shrinking since it had been formed.

(d) What new theory was developed using Wegener's ideas?

    A   convection currents

    B   plate tectonics

    C   radioactive decay

    D   seismic activity

(4 marks)

Chemistry B

1 Read the information in the box and use it to help you to answer the questions.

See pages 76–7

> Smart polymers can be used to switch enzymes on and off. The polymers are described as 'smart' because they alter their properties when conditions such as light, temperature or acidity change. Tiny smart polymer chains are attached to enzymes next to the active sites. The chains extend or contract depending on the conditions and block or unblock the site. This switches the enzyme off or on. One application already in use is for drugs that need to remain inactive until they reach a particular place in the body.

(a) What is a polymer? *(2 marks)*

(b) Why are the polymers in the article 'smart'? *(2 marks)*

(c) Suggest another application for smart polymers that is not mentioned in the article. *(1 mark)*

2 Read the information in the box and use it to help you to answer the questions.

> Many plant oils are drying oils, which restricts their use as fuels. Drying happens when double bonds in the molecules are oxidised and form cross-links so the oil polymerises into a plastic-like solid. The process is accelerated at high temperatures and engines quickly become gummed up. Oils with high iodine values have more double bonds. An iodine value of less than 25 is required if the oil is to be used in unmodified diesel engines.

(a) What is a drying oil? *(2 marks)*

(b) In what way are the polymers formed like thermosetting plastics? *(1 mark)*

(c) Suggest why drying oils are used in oil paints. *(2 marks)*
See page 75

(d) Describe how you could find the volume of iodine solution that reacted with a plant oil. Include how you would make your results as precise as possible. *(4 marks)*
See pages 83, 95

(e) Iodine values can be lowered by hydrogenation.
See page 85

   (i) How could this be done? *(2 marks)*

   (ii) Give one disadvantage of hydrogenation. *(1 mark)*
See page 93

GET IT RIGHT!

Do not worry if the questions are about something you have not met before or something you have not studied during the course. The information that you need to answer the questions will be given on the paper and the questions will be testing your understanding. Read the information carefully and be sure to use it in your answers.

C2 | Additional chemistry

What you already know

Here is a quick reminder of previous work that you will find useful in this unit:

Diamond is made up only of carbon atoms

- Elements combine through chemical reactions to form compounds.
- We can represent compounds by formulae and we can summarise reactions by word equations and balanced symbol equations.
- Mass is conserved in chemical reactions because the same atoms are there before and after the reaction.
- Many, but not all, metals react with oxygen, water, acids.
- Displacement reactions can take place between metals and solutions of salts of other metals. Displacement can also occur between metals and metal oxides.
- Bases react with acids.
- We can identify patterns in chemical reactions.

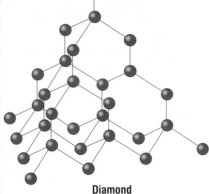 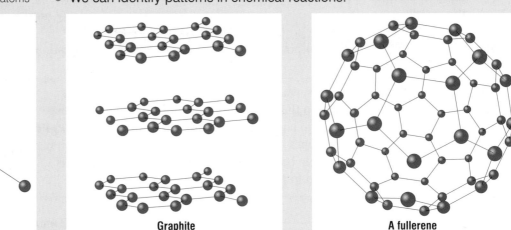

| Diamond | Graphite | A fullerene |

Diamond, graphite and the fullerenes are all forms of the element, carbon

RECAP QUESTIONS

1. Write a word equation for the reaction between iron and sulfur.
   How can we show that a new substance is formed in this reaction?

2. When 16 g of oxygen react with 2 g of hydrogen, how much water is formed?

3. Complete the word equation:

   calcium (metal) + hydrochloric acid → ......

4. How many types of atom are there in:
   a) sodium chloride (NaCl),
   b) ethanol (C_2H_5OH),
   c) sulfuric acid (H_2SO_4)?

5. How many different types of atoms are there in a pure sample of an element?

6. When we put zinc metal into blue copper sulfate solution a pink solid forms and the solution becomes much lighter in colour.
   What is happening?

7. When magnesium burns in air it produces a white powder. The magnesium gets very hot and you see a bright white light.
   Why do these observations suggest that a chemical reaction is taking place?

Making connections

Developing ideas about substances

Many people think that Antoine Lavoisier was the father of modern chemistry. If this is so, then his wife, Marie Anne may well be the mother. She was well educated, and translated documents and illustrated his scientific texts with great skill.

Antoine Lavoisier lived in France between 1743 and 1794. His experiments were some of the first proper chemical experiments involving careful measurements. For example, in chemical reactions he carefully weighed reactants and products. This was an important advance over the work of earlier chemists.

Working with his wife, Lavoisier showed that the quantity of matter is the same at the end as at the beginning of every chemical reaction. Working with other French chemists, Lavoisier invented a system of chemical names which described the structure of chemical compounds. Many of these names are still in use, including names such as sulfuric acid and sulfates.

ACTIVITY

The three scientists described here made enormous contributions to our understanding of the behaviour of matter and chemistry. Using this information, produce a poster with a timeline showing how our understanding of the behaviour of matter changed in the period from the early 18th century to the beginning of the 20th century.

You could research these ideas further using the Internet, especially at www.timelinescience.org.

Michael Faraday came from very humble beginnings, but his work on electricity and chemistry still affects our lives today. His achievements were acknowledged when his portrait was included on the £20 note in 1991.

Born in Yorkshire in 1791, Michael Faraday was one of 10 children. Apprenticed to a bookbinder, Faraday became an assistant to the great chemist Sir Humphry Davy. After hearing some of Davy's lectures in London, he sent him a bound copy of some notes he had made and was taken on.

After much work on electricity, Faraday turned his attention to electrolysis. He produced an explanation of what happens when we use an electric current to split up a chemical compound. Not only did Faraday explain what happens, he also introduced the words we still use today – *electrolysis*, *electrolyte* and *electrode*.

Ernest Rutherford was born in New Zealand in 1871. After his education in New Zealand he worked and studied in England and Canada. Then in 1910 he showed that the structure of the atom consists of a tiny positively charged nucleus that makes up nearly all of the mass of the atom. The nucleus is surrounded by a vast space which contains the electrons – but most of the atom is simply empty space! Rutherford received the Nobel Prize for Chemistry in recognition of his huge contribution to our understanding of the atom.

Ernest Rutherford was responsible for producing the evidence that completely changed our ideas about the structure of atoms

Chapters in this unit

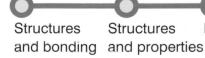
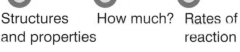

Structures and bonding — Structures and properties — How much? — Rates of reaction — Energy and reactions — Electrolysis — Acids, alkalis and salts

C2 1.1 Atomic structure

LEARNING OBJECTIVES

1 What is inside atoms?
2 Why is the number of protons in an atom equal to the number of electrons?
3 What is the order in which atoms are arranged in the periodic table?
4 What is the charge on a proton, neutron and electron?

In the middle of an atom is a small nucleus. This contains two types of particles, which we call **protons** and **neutrons**. A third type of particle orbits the nucleus – we call these particles **electrons**. Any atom has the same number of electrons orbiting its nucleus as it has protons in its nucleus.

Protons have a positive charge while neutrons have no charge – they are neutral. So the nucleus itself has an overall positive charge.

The electrons orbiting the nucleus are negatively charged. The size of the negative charge on an electron is exactly the same as the size of the positive charge on a proton. (In other words, the relative charge on a proton is $+1$, while the relative charge on an electron is -1.)

Because any atom contains equal numbers of protons and electrons, the overall charge on any atom is exactly zero. For example, a carbon atom has 6 protons, so we know it also has 6 electrons.

a) What are the names of the three particles that make up an atom?
b) An oxygen atom has 8 protons – how many electrons does it have?

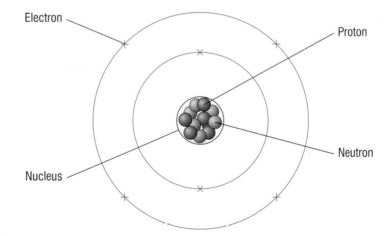

Figure 1 Understanding the structure of an atom gives us important clues to the way chemicals react together

Type of sub-atomic particle	Relative charge
proton	$+1$
neutron	0
electron	-1

To help you remember the charge on the sub-atomic particles:

Protons are **P**ositive;
Neutrons are **N**eutral;
so that means **E**lectrons must be **N**egative!

Atomic number

We call the number of protons in the nucleus of an atom its **atomic number** or **proton number**.

As all of the atoms of a particular element have the same number of protons, they also have the same atomic number. So the atomic number of hydrogen is 1 and it has one proton in the nucleus. The atomic number of carbon is 6 and it has 6 protons in the nucleus. The atomic number of sodium is 11 and it has 11 protons in the nucleus.

Each element has its own atomic number. If you are told that the atomic number of an element is 8, you can identify that element from the periodic table. In this case it is oxygen.

c) Which element has an atomic number of 14?

Elements in the periodic table are arranged in order of their atomic numbers.

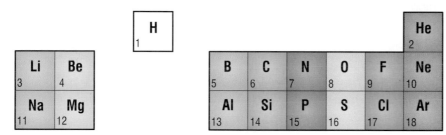

Figure 2 The elements in the periodic table are arranged in order of their atomic numbers

You read the periodic table from left to right, and from the top down – just like reading a page of writing.

Look at the atomic numbers of the elements in the last group of the periodic table:

d) What do you notice about the atomic numbers going from helium to neon to argon?

You will be able to explain this pattern when you learn more about the arrangement of electrons in atoms later in this chapter.

DID YOU KNOW?

In 1808, a chemist called John Dalton published a theory of atoms, explaining how these joined together to form new substances. Not everyone liked his theory though – one person wrote 'Atoms are round bits of wood invented by Mr Dalton!'

SUMMARY QUESTIONS

1 Copy and complete using the words below:

atomic   electrons   negative   neutrons   protons

In the nucleus of atoms there are ...... and .... Around the nucleus there are ...... which have a ...... charge. In the periodic table, atoms are arranged in order of their ...... number.

2 Use the periodic table in Figure 2 to find the atomic number of the elements lithium, sulfur, magnesium, chlorine and nitrogen.

3 Atoms are always neutral. Explain why this means that an atom must always contain the same number of protons and electrons.

KEY POINTS

1 Atoms are made of protons, neutrons and electrons.
2 Protons and electrons have equal and opposite electric charges. Protons are positively charged, and electrons are negatively charged.
3 Atoms are arranged in the periodic table in order of their atomic number.
4 Neutrons have no electric charge. They are neutral.

C2 1.2

The arrangement of electrons in atoms

LEARNING OBJECTIVES

1 How are the electrons arranged inside an atom?
2 How is the number of electrons in the highest energy level of an atom related to its group in the periodic table?
3 How is the number of electrons in the highest energy level of an atom related to its chemical properties?

GET IT RIGHT!

Make sure that you can draw the electronic structure of the atoms of all of the first 20 elements when you are given their atomic numbers.

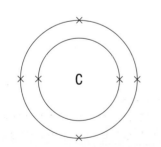

Figure 2 A simple way of representing the electrons in a carbon atom and the energy levels where they are found. We can show this as 2,4. This is called the **electronic structure** (or **electronic configuration**) of the atom.

One model of the atom which we use has electrons arranged around the nucleus in **shells**, rather like the layers of an onion. Each shell represents a different **energy level**. The lowest energy level is shown by the shell which is nearest to the nucleus.

With their negative charge, electrons are attracted to the positively charged nucleus. To move an electron from a shell close to the nucleus to one further away we need to put energy into the atom. The energy is needed to overcome this attractive force. This means that electrons in shells further away from the nucleus have more energy than electrons in shells closer to the nucleus.

a) Where are the electrons in an atom?
b) Which shell represents the lowest energy level in an atom?

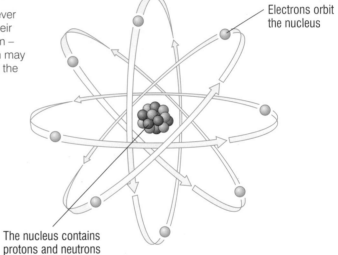

Figure 1 No-one has ever seen the electrons in their energy levels in an atom – this is one model which may help you to understand the structure of atoms

Electrons orbit the nucleus

The nucleus contains protons and neutrons

We could not possibly draw atoms which look like this every time we wanted to show the structure of an atom. It's easier to draw atoms as in Figure 2.

An energy level can only hold a certain number of electrons. The first, and lowest, energy level holds two electrons. The second energy level is filled up by eight electrons. Once there are eight electrons in the third energy level, the fourth begins to fill up, and so on.

Elements whose atoms have a full outer energy level are very stable and unreactive. They are called the **noble gases** – helium, neon and argon are examples.

The most usual way of drawing the arrangement of electrons in an atom is shown in Figure 2. We can also write down the numbers of electrons in each energy level.

The atomic number of an element tells us how many electrons there are in its atoms. For example, for the carbon atom in Figure 2 the atomic number is 6, giving us 6 electrons. This means that we write its **electronic structure** as 2,4.

An atom with the atomic number 13 has an electronic structure 2,8,3. This represents 2 electrons in the first, and lowest, energy level, then eight in the next energy level and 3 in the highest energy level (its outermost shell).

The best way to understand these arrangements is to look at some examples.

c) How many electrons can the first energy level hold?

Filling up the energy levels (shells)

We call the horizontal rows of the periodic table **periods**. As we move across a period of the table, each element has one more electron in its highest energy level (or outer shell) than the element before it. When we start a new period, a new energy level begins to fill with electrons.

The pattern is quite complicated after argon. However, the elements in the main groups all have the same number of electrons in their highest energy level. These electrons are often called the outer electrons because they are in the outer shell.

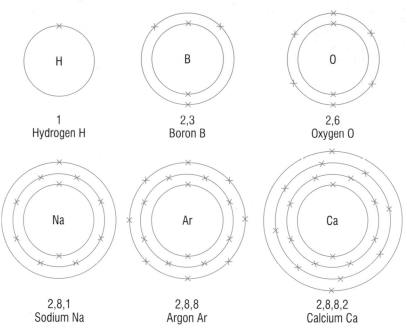

| 1 | 2,3 | 2,6 |
| Hydrogen H | Boron B | Oxygen O |

| 2,8,1 | 2,8,8 | 2,8,8,2 |
| Sodium Na | Argon Ar | Calcium Ca |

Figure 3 Once you know the pattern, you should be able to draw the energy levels and electrons in any of the first 20 atoms (given their atomic number)

All the elements in Group 1 have one electron in their highest energy level and the noble gases, except for helium, have 8 electrons in their highest energy level.

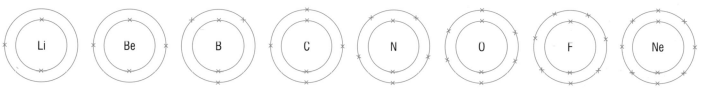

We call the vertical columns of the periodic table **groups**. The chemical properties of an element depend on how many electrons it has. Most importantly, the way an element reacts is determined by the number of electrons in its highest energy level or outer shell. As we have seen, the elements in a particular group all have the same number of electrons in their highest energy levels. This means that they all share similar chemical properties.

Figure 4 As a period builds up, the number of electrons in the outer shell of each element increases by one

---

SUMMARY QUESTIONS

1 Copy and complete using the words below:

   **electron   energy   energy levels   group   nucleus
   period   shells**

   The electrons in an atom are arranged around the …… in …… or …… …… . The electrons further away from the nucleus have more …… than those close to the nucleus. As you go across a …… of the periodic table, each element has one more …… than the previous element. All elements in the same …… have the same number of electrons in their outer shell.

2 Draw the arrangement of electrons in the following atoms:

   a) Li   b) B   c) P   d) Ar.

3 What is special about the electronic structure of neon and argon?

KEY POINTS

1 The electrons in an atom are arranged in energy levels or shells.
2 Atoms with the same number of electrons in their outer shell belong in the same group of the periodic table.
3 The number of electrons in the outer shell of an atom determines the way that the atom behaves in chemical reactions.

C2 1.3 Chemical bonding

LEARNING OBJECTIVES

1. How do elements form compounds?
2. Why do the elements in Group 1 react with the elements in Group 7?

You already know that we can mix two substances together without either of them changing. For example, we can mix sand and salt together and then separate them again. No change will have taken place. We can even dissolve sugar in tea and separate it out again. But in chemical reactions the situation is very different.

When the atoms of two or more elements react they make a compound. The compound formed is different to both of them and we cannot get either of the elements back again easily. We can also react compounds together to form other compounds, but the reaction of elements is easier to understand as a starting point.

a) What is the difference between **mixing** two substances and **reacting** them?

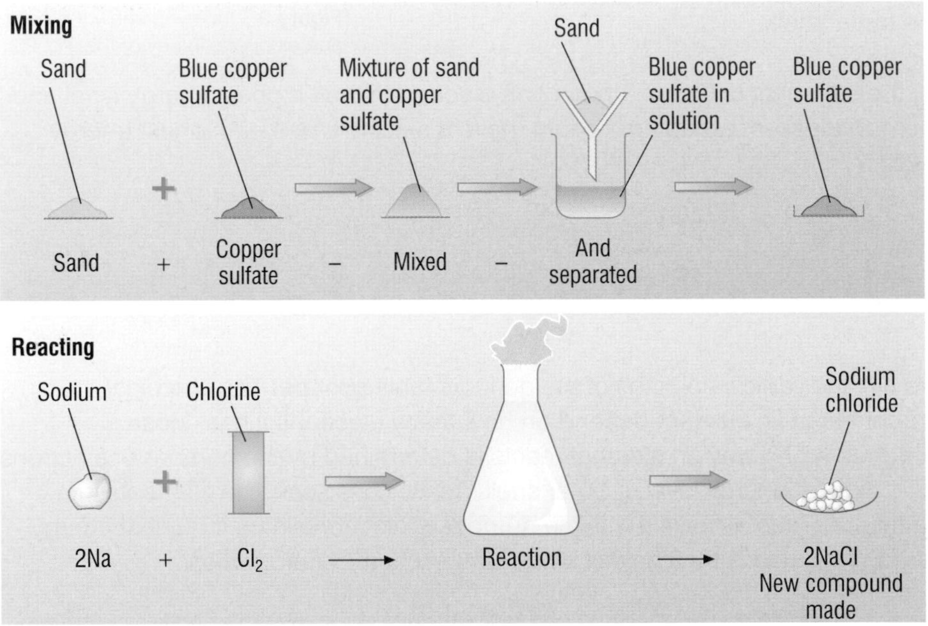

Figure 1 The difference between mixing and reacting. Separating mixtures is usually quite easy, but separating substances once they have reacted can be quite difficult.

Why do atoms react?

When an atom has a full outer shell it is stable and unreactive (like the noble gases in Group 0). However most atoms do not have a full outer shell. When atoms react they take part in changes which give them a stable arrangement of electrons. They may do this by either:

- sharing electrons, which we call **covalent bonding**, or by
- transferring electrons, which we call **ionic bonding**.

In ionic bonding the atoms involved lose or gain electrons so that they have a noble gas structure. So for example, if sodium, 2,8,1 loses one electron it is left with the stable electronic structure of neon 2,8.

However, it is also left with one more proton in the nucleus than there are electrons in orbit around the nucleus. The proton has a positive charge so the sodium atom has now become a positively charged particle. We call this a **sodium ion**. The sodium ion has a single positive charge. We write the formula of a sodium ion as Na^+. The electronic structure of the Na^+ ion is $[2,8]^+$.

b) When atoms join together by **sharing** electrons, what type of bond is this?

c) When atoms join together as a result of **gaining** or **losing** electrons, what type of bond is this?

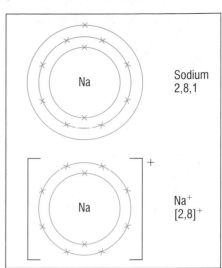

Figure 2 A positive sodium ion (Na⁺) is formed when a sodium atom loses an electron during ionic bonding with another element

Similarly some atoms gain electrons during reactions to achieve a stable noble gas structure. Chlorine, for example, has the electronic structure 2,8,7. By gaining a single electron, it gets the stable electronic structure of argon [2,8,8].

In this case there is now one more electron than there are positive protons in the nucleus. So the chlorine atom becomes a negatively charged particle known as a **chloride ion**. This carries a single negative charge. We write the formula of the chloride ion as Cl^-. Its electronic structure is $[2,8,8]^-$.

Representing ionic bonding

When atoms react together to form ionic bonds, atoms which need to lose electrons react with elements which need to gain electrons. So when sodium reacts with chlorine, sodium loses an electron and chlorine gains that electron so they both form stable ions.

We can show this in a diagram. Look at Figure 4:

The electrons of one atom are represented by dots, and the electrons of the other atom are represented by crosses.

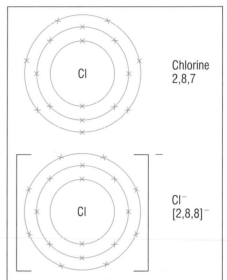

Figure 3 A negative chloride ion (Cl⁻) is formed when a chlorine atom gains an electron during ionic bonding with another element

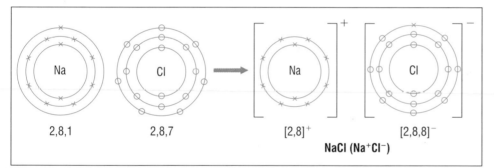

2,8,1   2,8,7 $[2,8]^+$ $[2,8,8]^-$

NaCl (Na⁺Cl⁻)

Figure 4 The formation of sodium chloride (NaCl) – an example of ion formation by transferring a single electron

SUMMARY QUESTIONS

1 Copy and complete using the words below:

    covalent   difficult   gaining   ionic   losing   new
        noble   sharing

When two substances react together they make a …… substance and it is …… to separate them. Some atoms react by …… electrons – we call this …… bonding. Other atoms react by …… or …… electrons – we call this …… bonding. When atoms react in this way they tend to get the electronic structure of a …… gas.

2 Draw diagrams to show the ions that would be formed when the following atoms are involved in ionic bonding. For each one, state whether electrons have been lost or gained and show the charge on the ions formed.

a) aluminium (Al)   b) fluorine (F)
c) potassium (K)   d) oxygen (O)

KEY POINTS

1 Elements react to form compounds by gaining or losing electrons or by sharing electrons.

2 The elements in Group 1 react with the elements in Group 7 because Group 1 elements can lose an electron to gain a full outer shell. This electron can be given to an atom from Group 7, which then also gains a full outer shell.

C2 1.4   Ionic bonding

LEARNING OBJECTIVES

1 How are ionic compounds held together?
2 Which elements, other than those in Groups 1 and 7, form ions?

You have seen how positive and negative ions form during some reactions. Ionic compounds are usually formed when metals react with non-metals.

The ions formed are held to each other by enormously strong forces of attraction between the oppositely charged ions. This electrostatic force of attraction, which acts in all directions, is called the **ionic bond**.

The ionic bonds between the charged particles results in an arrangement of ions that we call a **giant structure**. If we could stand among the ions they would seem to go on in all directions for ever.

The force exerted by an ion on the other ions in the lattice acts equally in all directions. This is why the ions in a giant structure are held together so strongly.

The giant structure of ionic compounds is very *regular*. This is because the ions all pack together neatly, like marbles in a tin or apples in a box.

a) What name do we give to the arrangement of ions in an ionic compound?
b) What holds the ions together in this structure?

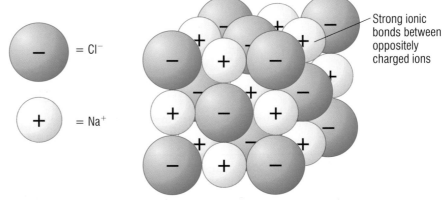

= Cl^-

= Na^+

Strong ionic bonds between oppositely charged ions

Figure 1 A giant ionic lattice (3D network) of sodium and chloride ions

DID YOU KNOW?

Common salt is sodium chloride. In just 58.5 g of salt there are over 600 000 000 000 000 000 000 000 ions of Na^+ and the same number of Cl^- ions.

Other ionic compounds

Sometimes the atoms reacting need to gain or lose two electrons to gain a stable noble gas structure. An example is when magnesium (2,8,2) reacts with oxygen (2,6). When these two elements react they form magnesium oxide (MgO). This is made up of magnesium ions with a double positive charge (Mg^{2+}) and oxide ions with a double negative charge (O^{2-}).

We can represent the atoms and ions involved in forming ions by *dot and cross diagrams*. In these diagrams we only show the electrons in the outermost shell of each atom or ion. So they are quicker to draw than the diagrams on the previous page. Look at Figure 2 on the next page:

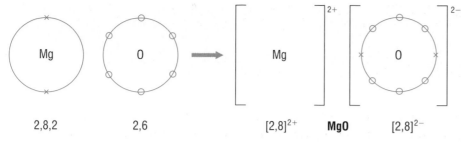

2,8,2   2,6 $[2,8]^{2+}$ **MgO** $[2,8]^{2-}$

Figure 2 When magnesium oxide (MgO) is formed the reacting atoms lose or gain two electrons

In some cases one of the atoms needs to gain or lose more electrons than the other has to lose or gain. In this case, two or more atoms of each element may react.

For example, think about calcium chloride. Each calcium atom needs to lose two electrons but each chlorine atom needs to gain only one electron. This means that two chlorine atoms react with every one calcium atom to form calcium chloride. So the formula of calcium chloride is $CaCl_2$.

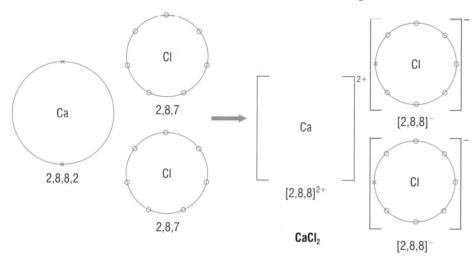

Figure 3 The formation of calcium chloride (CaCl₂)

SUMMARY QUESTIONS

1 Copy and complete the table:

Atomic number	Atom	Electronic structure of atom	Ion	Electronic structure of ion
8	O	c)	e)	$[2,8]^{2-}$
19	a)	2,8,8,1	K⁺	g)
17	Cl	d)	Cl⁻	h)
20	b)	2,8,8,2	f)	i)

j) Explain why potassium chloride is KCl but potassium oxide is K₂O.
k) Explain why calcium oxide is CaO but calcium chloride is CaCl₂.

2 Draw dot and cross diagrams to show how you would expect the following elements to form ions together:

a) lithium and chlorine,
b) calcium and oxygen,
c) aluminium and chlorine.

KEY POINTS

1 Ionic compounds are held together by strong forces between the oppositely charged ions. This is called ionic bonding.
2 Other elements that can form ionic compounds include those in Groups 2 and 6.

121

C2 1.5   Covalent bonding

Figure 1 Many of the substances which make up the living world are held together by covalent bonds between non-metal atoms

Reactions between metals and non-metals usually result in ionic bonding. However many, many compounds are formed in a very different way. When non-metals react together they share electrons to form molecules. We call this **covalent bonding**.

Simple molecules

The atoms of non-metals generally need to gain electrons to achieve stable outer energy levels. When they react together neither atom can give away electrons, so they get the electronic structure of a noble gas by sharing electrons. The atoms in the molecules are then held together because they are sharing pairs of electrons. We call these strong bonds between the atoms **covalent bonds**.

a) What is the bond called when two atoms share electrons?

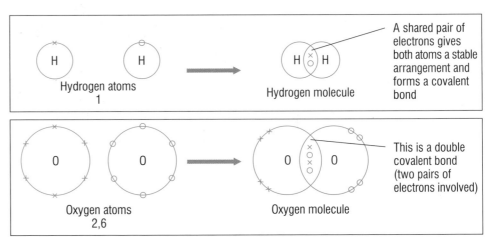

Figure 2 Atoms of hydrogen and oxygen join together to form stable molecules in which the atoms are held together by covalent bonds

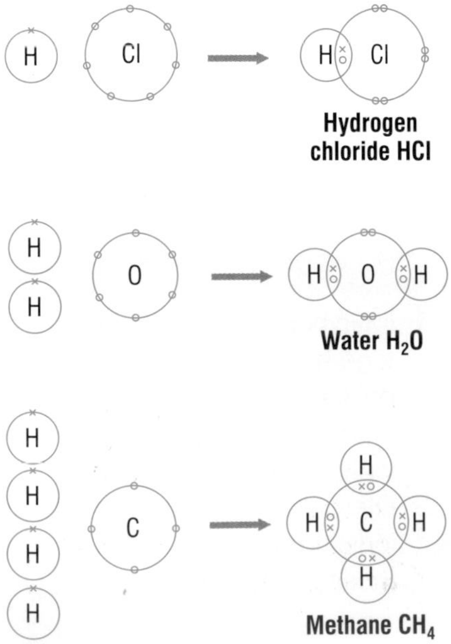

Sometimes in covalent bonding each atom brings the same number of electrons to the reaction for sharing. But this is not always the case. Sometimes one element will need several electrons, while the other element only needs one more electron for a stable arrangement. In this case, more atoms become involved in the reaction.

b) How many electrons are shared in a covalent bond?

We can represent the covalent bonds in substances such as water, ammonia and methane in a number of ways. Each way of representing them means exactly the same thing – it just depends on what we want to show.

Hydrogen chloride HCl

Water H₂O

Methane CH₄

Figure 3 The principles of covalent bonding remain the same however many atoms are involved

Water
H₂O

Figure 4 We can represent a covalent compound by showing the highest energy level, the outer electrons or just the fact that there are a certain number of covalent bonds

Giant structures

Many substances containing covalent bonds consist of small molecules, for example, H₂O. However some covalently bonded substances are very different. They have giant structures where huge numbers of atoms are held together by a network of covalent bonds.

Diamonds have a giant covalent structure. In diamond, each carbon atom forms four covalent bonds with its neighbours in a rigid giant covalent lattice.

Silicon dioxide (silica) is another substance with a giant covalent structure.

c) What do we call the structure of a substance held together by a network of covalent bonds?

NEXT TIME YOU...

... see a diamond ring, think about what properties make the diamond suited to its purpose.

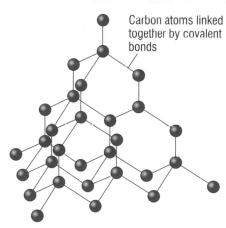

Carbon atoms linked together by covalent bonds

Figure 5 Part of the giant covalent structure of diamond

Figure 6 Diamonds owe their hardness and long-lasting nature to the way the carbon atoms are arranged

SUMMARY QUESTIONS

1 Copy and complete using the words below:

   covalent   giant   molecules   shared

   When non-metal atoms react together they tend to produce ...... bonds. The atoms in these bonds are held together by ...... electrons. Most substances held together by covalent bonds consist of ......, but a few have ...... structures.

2 Draw diagrams to show the covalent bonds between the following atoms.

   a) two hydrogen atoms
   b) two chlorine atoms
   c) a hydrogen atom and a fluorine atom

3 Draw dot and cross diagrams to show the covalent bonds when:

   a) a nitrogen atom bonds with three hydrogen atoms
   b) a carbon atom bonds with two oxygen atoms.

KEY POINTS

1 Covalent bonds are formed when atoms share electrons.
2 Many substances containing covalent bonds consist of molecules, but some have giant covalent structures.

C2 1.6   Bonding in metals

LEARNING OBJECTIVES

1 How are the atoms in metals arranged?
2 How are the atoms in metals held together? [Higher]

Metals are another example of giant structures. You can think of metal as a lattice of metal atoms (or positively charged ions), arranged in regular layers.

The outer electrons in each atom can easily move from one atom to the next one. The outer electrons (in the highest occupied energy level) form a 'sea' of free electrons surrounding positively charged metal ions. Strong electrostatic attraction between the negatively charged electrons and positively charged ions bond the metal ions together. The electrons act a bit like a glue!

a) Which electrons do metal atoms use to form metallic bonds?

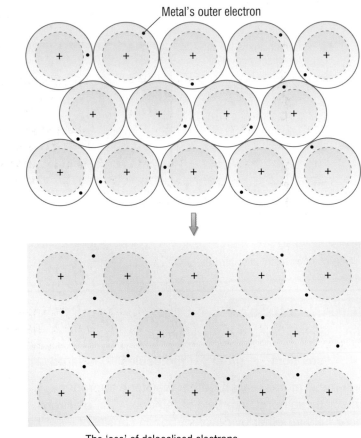

Metal's outer electron

The 'sea' of delocalised electrons

Figure 1 A metal consists of positively charged metal ions surrounded by a 'sea' of electrons

The 'sea' of free electrons are called **delocalised electrons**. These electrons help us explain the properties of metals.

b) What do you think happens to the delocalised electrons when an electric current flows through a metal?

Metal crystals

The giant structure of a metal is not usually the same all through the metal. If you look very closely at a metal surface that has been specially prepared, you can see that the metal is made up of a number of small crystals. We call these **grains**, and the places where they join are the **grain boundaries**.

c) What do we call the crystals in metals?

PRACTICAL

Growing silver crystals

You can grow crystals of silver metal by suspending a length of copper wire in silver nitrate solution. The crystals of silver will appear on the wire quite quickly, but for best results they need to be left for several hours!

— Copper wire

— Boiling tube containing silver nitrate solution

Figure 2 Growing silver crystals

- Explain your observations.

Sometimes you can see metal crystals on the surface of steel that has been dipped in zinc to prevent it from rusting. We call this treatment **galvanising**. Galvanised steel is used to make channels for carrying insulated electric wires and many other containers, and is not usually painted.

d) Why do we galvanise steel?

PRACTICAL

Survey of metallic crystals

Take a look round your school to see if you can find any galvanised steel. See if you can spot the metal crystals. You can also look for crystals on brass fittings that have been left outside and not polished.

SUMMARY QUESTIONS

1 Copy and complete using the words below:

electrons   electrostatic   free   giant   outer   positive

Metals also have ...... structures. The atoms in these are held together by ...... from the ...... energy levels of the metal atoms. The ...... ions that this produces are held together by strong ...... forces. The electrons in metals are ...... to move throughout the structure.

2 Find out how we can change the grain size in a metal and how this affects its properties.

3 Explain why the electrons in a metal are both like glue and not like glue. [Higher]

KEY POINTS

1 The atoms (or ions) in metals are arranged in regular layers.
2 The positive ions in metals are held together by electrons from the outer shell of each metal atom. These delocalised electrons are free to move throughout the metal lattice. [Higher]

C2 1.7   The history of the atom

John Dalton's atomic theory

John Dalton – the man who gave us atoms

John Dalton was born in 1766 in the Lake District in England. His father was a weaver who taught John at home before sending him to a Quaker school in Eaglesfield, where they lived. John was amazingly clever – by the time he was 12 he was teaching other children!

He was interested in almost everything. He made observations of the weather as well as being the first person to study colour-blindness. (John was colour-blind himself – see the photo below.)

But Dalton is best-remembered for his ideas about chemistry – and in particular his theories about atoms. As a result of a great deal of work, Dalton suggested that:

- All matter is made up of indivisible particles called atoms.
- Atoms of the same element are similar in mass and shape but differ from the atoms of other elements.
- Atoms cannot be created or destroyed.
- Atoms join together to form compound atoms (what we would now call molecules) in simple ratios.

Dalton's statements were backed up with much research, even though not all of it was accurate. For example, he insisted that one hydrogen atom combined with one oxygen atom to form water. However, most of his research reflected the same results as other scientists of the time were getting.

Dalton's atomic theory explained much of what scientists were seeing, and so his idea of atoms was accepted relatively quickly. Some scientists even made wooden models of atoms of different elements, to show their different relative sizes.

By 1850, the atomic theory of matter was almost universally accepted and virtually all opposition had disappeared. Dalton's atomic theory was the basis of much of the chemistry done in the rest of the 19th and early 20th centuries.

John Dalton's eyes (on the watch-glass) were taken out after his death as he requested. He wanted a doctor to check his theory of colour blindness. Unfortunately this theory proved incorrect.

ACTIVITY

Imagine that you are John Dalton and that you have just finished writing a book about your ideas on atoms. Write a letter to someone explaining your ideas. You can choose to write to:

- another scientist,
- a member of your family,
- a journalist who is interested in your ideas and who wants to know more about them in order to write a newspaper article for the general public.

Atoms and the future

Deep underneath the Swiss countryside lies a huge maze of tunnels. Inside these tunnels, scientists are working to puzzle out the structure of the atom. They are searching for the particles that make up the protons and neutrons inside each atom.

To find these tiny particles they need to use huge machines. These accelerate particles like electrons and protons up to speeds close to the speed of light. Then the particles smash into each other in a kind of 'subatomic demolition derby'!

This is a particle detector under construction

It's really important that we know as much as we can about atoms. Although it doesn't seem like this knowledge is very useful at the moment, it could lead to important discoveries in the future. And besides, we should try to find out as much as we can about the world around us!

This shows a section of a particle accelerator

The money that's spent on this kind of research is enormous. We should spend money on APPLIED kinds of scientific research that may be able to help people, not on research that isn't any practical use.

ACTIVITY

Research like this costs a great deal of money. Who do you agree with?

Design a poster to show your ideas.

SUMMARY QUESTIONS

1 a) Unscramble the following words to make the names of the three different particles in an atom:

nropto   erontun   lentroce

b) Now show the charge on each of these particles by writing one of the following words next to each name – neutral, positive, negative.

2 Draw the structure of the following atoms showing all the energy levels in each atom:

a) helium (He, atomic number 2),

b) oxygen (O, atomic number 8),

c) potassium (K, atomic number 19),

d) chlorine (Cl, atomic number 17),

e) aluminium (Al, atomic number 13).

3 The diagrams show the energy levels in three atoms: (The letters are NOT the chemical symbols.)

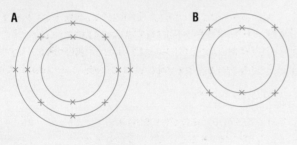

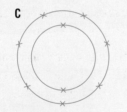

a) Which atom belongs to group 2?

b) To which group does atom C belong?

c) Atom B bonds with four atoms of hydrogen. Draw a dot and cross diagram to show the compound that is formed.

d) Draw dot and cross diagrams to show how atom A bonds with C atoms.

4 Describe, with diagrams, how the particles are held together in the following substances:

a) a molecule of bromine (Br_2),

b) a sample of diamond (carbon).

c) a salt crystal (NaCl).

5 Explain the bonding in sodium metal. You may wish to include a diagram. (The atomic number of sodium is 11.)   [Higher]

EXAM-STYLE QUESTIONS

1 The diagram represents an atom of an element.

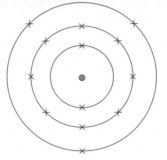

(a) Write the electronic structure of this atom as numbers and commas.   (1)

(b) How many protons are in the nucleus of this atom?   (1)

(c) Name the other particles that are in the nucleus. (1)

(d) In which group of the periodic table is this element?   (1)

(e) Draw a similar diagram to show the ion formed by this atom in ionic compounds. Show the charge on the ion.   (2)

2 Complete the missing information (a) to (f) in the table.

Atomic number	Symbol	Electronic structure of atom	Formula of ion	Electronic structure of ion
9	F	(a)	(b)	$[2,8]^-$
11	(c)	2,8,1	Na^+	(d)
(e)	S	2,8,6	S^{2-}	(f)

(6)

3 A hydrogen atom can be represented by the diagram:

(a) Draw a similar diagram to show the electrons in the outer shell of a chlorine atom.   (1)

(b) Draw a dot and cross diagram to show the bonding in a molecule of hydrogen chloride.   (2)

(c) Explain why hydrogen and chlorine form a single covalent bond.   (2)

(d) Explain why silicon can form giant structures.   (3)

4 (a) Draw a dot and cross diagram to show the arrangement of electrons in a magnesium ion. Show the charge on the ion. (3)

(b) Draw a dot and cross diagram to show the arrangement of electrons in an oxide ion. (3)

(c) What is the formula of magnesium oxide? (1)

5 Berzelius (1779–1848) carried out experiments to discover the atomic mass of many elements. He wrote about the fact that bodies combine in definite proportions and that led him to suggest the existence of a cause.

(a) Suggest an observation that Berzelius might have made. (1)

(b) Is what Berzelius wrote a prediction or a hypothesis? Explain your answer. (1)

(c) Berzelius gave oxygen the number 100 to represent its relative atomic mass. He then set out to compare the mass of other elements with oxygen. However, he could not measure these directly because they could not be turned into gases – the temperature needed was too high and he did not have the equipment to do this.

(i) Explain, in general terms, the problem he had. (1)

(ii) Use this example to explain the relationship between technology and science. (1)

6 The diagram represents atoms of potassium in the solid metal.

(a) What is the electronic structure of a potassium atom? (1)

(b) Explain as fully as you can how the atoms are held together in solid potassium metal. (3)

[Higher]

HOW SCIENCE WORKS QUESTIONS

How the atomic theory was developed

2,500 years ago, Democritus believed that matter could be broken into smaller and smaller pieces until finally there would be particles that were 'indivisible' – the Greek word for this is *atomos*. He thought they looked like this:

Humphry Davy, who went to Truro Grammar School, discovered many of the elements that we are familiar with in chemistry lessons. He separated potassium, sodium and chlorine. As he couldn't break these elements down any further, he said that this must be the definition of an element.

Dalton became convinced that each element was made of a different kind of atom. He can be credited with the first scientific use of the term 'atom', although the Greeks had used the idea thousands of years before.

Dalton believed that

- the atom must be very small,
- all matter is made from atoms, and
- these atoms cannot be destroyed.

He gave hydrogen the atomic weight of 1, because he knew it to be the lightest atom.

He thought water was made of 1 hydrogen and 1 oxygen atom and therefore predicted that oxygen must have an atomic weight of 7.

Berzelius, a Swedish chemist, tested Dalton's theory experimentally. He correctly found the atomic weights of 40 elements.

a) When was the first theory of the atom put forward? (1)

b) What observation led to the definition of an element? (1)

c) What hypothesis did Dalton come up with? (1)

d) What prediction was made by Dalton? (1)

e) Check in the periodic table whether Dalton's prediction was correct. (1)

f) What was Berzelius' contribution to the atomic theory? (1)

g) Is Dalton's atomic theory completely true? Explain your answer. (1)

C2 2.1   Ionic compounds

We have already seen that an ionic compound consists of a giant structure of ions arranged in a lattice. The attractive electrostatic forces between the oppositely charged ions act in all directions and they are also very strong. This holds the ions in the lattice together very tightly.

a) What type of force holds the ions together in an ionic compound?

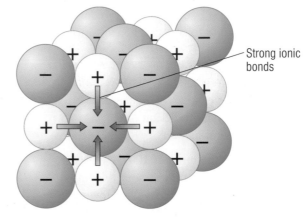

Figure 1 The attractive forces between the oppositely charged ions in an ionic compound are very strong

Because the attractive forces between the oppositely charged ions in the lattice are very strong, and there are lots of them to overcome, it takes a lot of energy to break the lattice apart. This means that ionic compounds have high melting points and boiling points. Look at the graph in Figure 2.

b) Why do ionic compounds have high melting points and boiling points?

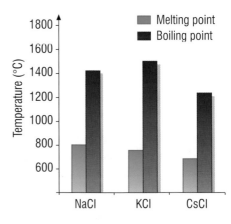

Figure 2 The strong attractive forces in a lattice of ions mean that ionic compounds have high melting points and boiling points

Once we have supplied enough energy to separate the ions from the lattice, they are free to move, and the ionic solid becomes a liquid. The ions are free to move anywhere in this liquid, so they are able to carry electrical charge through the molten liquid. A solid ionic compound cannot conduct electricity like this, because each ion is held in a fixed position in the lattice and cannot move around. They can only vibrate in their fixed positions.

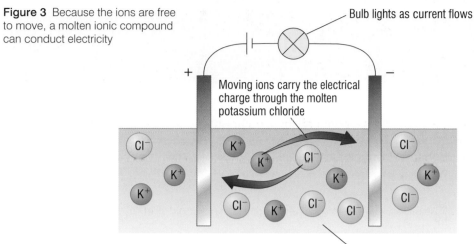

Figure 3 Because the ions are free to move, a molten ionic compound can conduct electricity

Bulb lights as current flows

Moving ions carry the electrical charge through the molten potassium chloride

Molten potassium chloride

Many ionic compounds will dissolve in water. When we dissolve an ionic compound in water the lattice is split up by the water molecules, and the ions are free to move. In the same way as molten ionic compounds will conduct electricity, solutions of ionic compounds will also conduct electricity. The ions in the solution are able to move around.

c) Why can ionic compounds conduct electricity when they are molten or dissolved in water?

Ionic solid	Molten ionic compound	Ionic compound in solution
Ions fixed in lattice – does not conduct electricity	High temperature provides energy to overcome strong attractive forces between ions. Ions free to move – will conduct electricity.	Water molecules separate ions from the lattice. Ions free to move – will conduct electricity.

PRACTICAL

Testing conductivity

Using a circuit as shown in Figure 3, dip a pair of electrodes into a 1 cm depth of sodium chloride crystals. What happens?

Now slowly add water.

● What happens?

● Explain your observations.

SUMMARY QUESTIONS

1 Copy and complete using the words below:

   attraction   conduct   high   lattice   molten   move
   oppositely   solution

   Ionic compounds have ...... melting points and boiling points because of the strong electrostatic forces of...... between ...... charged ions in the giant .... Ionic compounds will ...... electricity when ...... or in ...... because the ions are able to ...... freely.

2 Why is sea water a better conductor of electricity than water from a freshwater lake?

C2 2.2   Simple molecules

1 Which type of substances have low melting points and boiling points?
2 Why are some substances gases or liquids at room temperature? [Higher]
3 Why don't these substances conduct electricity?

When the atoms of non-metal elements react to form compounds, they share electrons in their outer shells. Then each atom gets a full outer shell of electrons. The bonds formed like this are called **covalent bonds**.

Water
H_2O

Figure 1 Covalent bonds hold the atoms found within molecules tightly together

a) How are covalent bonds formed?

Substances made up of covalently bonded molecules tend to have low melting points and boiling points.

Look at the graph in Figure 2.

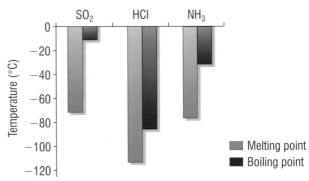

Figure 2 Substances made of simple molecules usually have low melting points and boiling points.

These low melting points and boiling points mean that many substances with simple molecules are liquids or gases at room temperature. Others are solids with quite low melting points, such as iodine and sulfur.

b) Do the compounds shown on the graph exist as solids, liquids or gases at 20°C?
c) You have a sample of ammonia (NH_3) at −120°C. Describe the changes that you would see as the temperature of the ammonia rises to 20°C (approximately room temperature).

Covalent bonds are very strong. So the atoms within each molecule are held very tightly together. However, each molecule tends to be quite separate from its neighbouring molecules. The attraction between the individual molecules in a covalent compound tends to be small. We say that there are weak *intermolecular forces* between molecules. Overcoming these forces does not take much energy.

HIGHER

d) How strong are the forces between the atoms in a covalent bond?
e) How strong are the forces between molecules in a covalent compound?

The covalent bonds between the hydrogen and oxygen atoms within a water molecule are strong. However, the forces of attraction between water molecules are relatively weak.

Look at the molecules in a sample of chlorine gas:

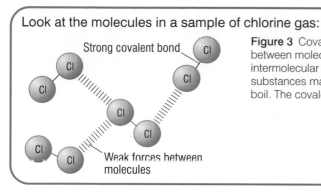

Strong covalent bond

Weak forces between molecules

Figure 3 Covalent bonds and the weak forces between molecules in chlorine gas. It is the weak intermolecular forces that are overcome when substances made of simple molecules melt or boil. The covalent bonds are **not** broken.

HIGHER

GET IT RIGHT!

Although the covalent bonds in molecules are strong, the forces between molecules are weak. [Higher]

Although a substance that is made up of simple molecules may be a liquid at room temperature, it will not conduct electricity.

Look at the demonstration below.

DEMONSTRATION

Conductivity

Figure 4 Compounds made of simple molecules do not conduct electricity

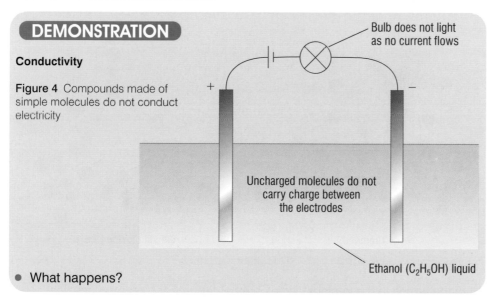

Bulb does not light as no current flows

Uncharged molecules do not carry charge between the electrodes

Ethanol (C_2H_5OH) liquid

● What happens?

Because there is no overall charge on the molecules in a compound like ethanol, the molecules cannot carry electrical charge. This makes it impossible for substances which are made up of simple molecules to conduct electricity.

f) Why don't molecular substances conduct electricity?

SUMMARY QUESTIONS

1 Copy and complete using the words below:

boiling   covalent   melting   molecules   strongly

Non-metals react to form ...... which are held together by ...... bonds. These hold the atoms together very .... The forces between molecules are relatively weak, so these substances have low ...... points and ...... points.

2 A compound called sulfur hexafluoride (SF_6) is used to stop sparks forming inside electrical switches designed to control large currents. Explain why the properties of this compound make it particularly useful in electrical switches.

3 The melting point of hydrogen chloride is −115°C whereas sodium chloride melts at 801°C. Explain why. [Higher]

KEY POINTS

1 Substances made up of simple molecules have low melting points and boiling points.
2 The forces between simple molecules are weak. These weak intermolecular forces explain their low melting points and boiling points. [Higher]
3 Simple molecules have no overall charge, so they cannot carry electrical charge. Therefore substances containing simple molecules do not conduct electricity.

C2 2.3   Giant covalent substances

LEARNING OBJECTIVES

1 How do substances with giant covalent structures behave?
2 Why is diamond hard and graphite slippery?
3 Why can graphite conduct electricity? [Higher]

While most non-metals react and form covalent bonds which join the atoms together in molecules, a few form very different structures. Instead of joining a small number of atoms together in individual molecules, the covalent bonds form large networks of covalent bonds. We call networks like this **giant covalent structures**. They are sometimes called macromolecules or giant molecular structures.

Substances such as diamond, graphite and silicon dioxide have giant covalent structures.

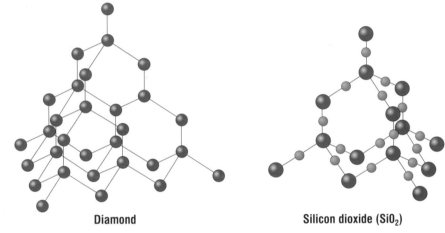

Diamond   **Silicon dioxide (SiO$_2$)**

Figure 1 The structures of diamond and silicon dioxide (sand)

All of the atoms in these giant lattices are held together by strong covalent bonds in both diamond and silicon dioxide. This gives these substances some very special properties. They are very hard, they have high melting points and boiling points and they are chemically very unreactive.

a) What do we call the structure of compounds which contain lots (millions) of atoms joined together by a network of covalent bonds?
b) What kind of physical properties do these substances have?

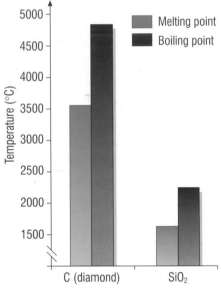

Figure 2 The large attractive forces in a giant lattice of covalently bonded atoms means that these compounds have high melting points and boiling points

Figure 3 Hard, shiny and transparent – diamonds make beautiful jewellery

We don't always find carbon as diamonds – another form is graphite (well known as the 'lead' in a pencil). In graphite, carbon atoms are arranged in giant layers. There are only weak forces between the layers so they can slide over each other quite easily.

c) Why is graphite slippery?

HIGHER

Another important property of graphite comes from the fact that there are free electrons within its structure. These free electrons allow graphite to conduct electricity, which diamond – and most other covalent compounds – simply cannot do. We call the free electrons found in graphite **delocalised electrons**. They behave rather like the electrons in a metallic structure.

The carbon atoms in graphite's layers are arranged in hexagons. So each carbon atom bonds to three others. (See Figure 4.) This leaves one spare outer electron on each carbon atom. It is this electron that becomes delocalised along the layers of carbon atoms.

d) Why can graphite conduct electricity?

Figure 4 The giant structure of graphite. When you write with a pencil, some layers of carbon atoms slide off the 'lead' and are left on the paper.

GET IT RIGHT!

Giant covalent structures are held together by covalent bonds throughout the structure.

Fullerenes

Apart from diamond and graphite, there are other different molecules that carbon can produce. In these structures the carbon atoms join together to make large cages which can have all sorts of weird shapes. Chemists have made shapes looking like balls, onions, tubes, doughnuts, corkscrews and cones!

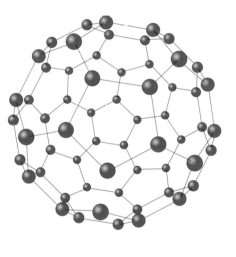

Chemists discovered carbon's ability to behave like this in 1985. We call the large carbon molecules containing these cage structures **fullerenes**. They are sure to become very important in nanoscience applications. (See pages 138 and 139.)

Figure 5 The first fullerene to be discovered contained only 60 carbon atoms, but chemists can now make *giant fullerenes* which contain many thousands of carbon atoms. Scientists can now place other molecules inside these carbon cages. This has exciting possibilities, including the delivery of drugs to specific parts of the body.

KEY POINTS

1 Some covalently bonded substances contain giant structures.
2 These substances have high melting points and boiling points.
3 The giant structure of graphite contains layers of atoms that can slide over each other which make graphite slippery. The atoms in diamond have a different structure and cannot slide like this – so diamond is a very hard substance.
4 Graphite can conduct electricity because of the delocalised electrons along its layers. [Higher]

SUMMARY QUESTIONS

1 Copy and complete using the words below:

   atoms   boiling   carbon   hard   high   layers   slide   soft

   Giant covalent structures contain many ...... joined by covalent bonds. They have ...... melting points and ...... points. Diamond is a very ...... substance because the ...... atoms in it are held strongly to each other. However, graphite is ...... because there are ...... of atoms which can ...... over each other.

2 Graphite is sometimes used to reduce the friction between two surfaces that are rubbing together. How does it do this?

3 Explain in detail why graphite can conduct electricity but diamond cannot. [Higher]

C2 2.4   Giant metallic structures

LEARNING OBJECTIVES

1 Why can we bend and shape metals?
2 Why do metals conduct electricity and heat? [Higher]

Figure 1 Drawing copper out into wires depends on being able to make the layers of metal atoms slide easily over each other

We can hammer and bend metals into different shapes, and draw them out into wires. This is because the layers of atoms in a pure metal are able to slide easily over each other.

Force

Atoms are all the same size — Pure metal

Layers slide over each other easily in a pure metal

Metal cooking utensils are used all over the world, because metals are good conductors of heat. Wherever electricity is generated, metal wires carry the electricity to where it is needed. That's because metals are also good conductors of electricity.

a) Why can metals be bent and shaped when forces are applied?

The atoms in metals are held together in a giant structure by a sea of delocalised electrons. These electrons are a bit like 'glue', holding the atoms (or positively charged ions) together. (See page 124.)

However, unlike glue the electrons are able to move throughout the whole lattice. Because they can move and hold the metal ions together at the same time, the delocalised electrons enable the lattice to distort so that the metal atoms can move past one another.

b) How are metal atoms held together?

Metals conduct heat and electricity as a direct result of the ability of the delocalised electrons to flow through the giant metallic lattice.

c) Why do metals conduct electricity and heat?

Figure 2 Metals are essential in our lives – the delocalised electrons mean that they are good conductors of both heat and electricity

PRACTICAL

Making models of metals

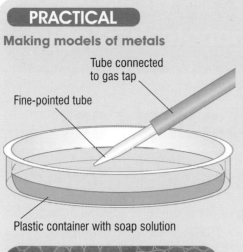

Tube connected to gas tap

Fine-pointed tube

Plastic container with soap solution

A regular arrangement of bubble 'atoms'

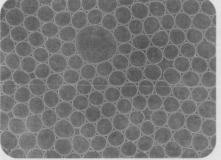

A larger bubble 'atom' has a big effect on the arrangement around it

Areas of bubble 'atoms' meet like grain boundaries within a metal

NEXT TIME YOU...

. . . get in a car, ride your bike or use anything made of metal, think how the metal object you are using has been made from a piece of metal with a very different shape. The fact that you can use it depends on the way that the layers of metal atoms can be persuaded to slide over each other!

We can make a model of the structure of a metal by blowing small bubbles on the surface of soap solution to represent atoms. Compressing or stretching the raft slightly leads to bubble 'atoms' being squashed together or pulled apart slightly. This shows how metals can return to their original shape after they have been bent slightly.

Compressing or stretching the bubble 'atoms' more leads to a permanent change in their position. This is what happens when we change the shape of a piece of metal permanently. In some areas a regular arrangement of bubble 'atoms' may be affected by a larger or smaller bubble. In others, areas of bubbles meet at different angles like the grain boundaries found in metals.

● Why are models useful in science?

SUMMARY QUESTIONS

1 Copy and complete using the words below:

delocalised   electricity   heat   shape   slide

The atoms in metals are held together by ...... electrons. These also allow the atoms to ...... over each other so that the metal's ...... can be changed. They also allow the metal to conduct ...... and ....

[Higher]

2 Use your knowledge of metal structures to explain how adding larger metal atoms to a metallic lattice can make the metal harder.

3 How can metals be hard and easily bent at the same time?

4 Explain why metals are good conductors of heat and electricity. [Higher]

KEY POINTS

1 We can bend and shape metals because the layers of atoms (or ions) in a metal can slide over each other.

2 Delocalised electrons in metals allow them to conduct heat and electricity well. [Higher]

C2 2.5   Nanoscience and nanotechnology

The science of tiny things – what can we do?

Nanoscience

Nanoscience is a new and exciting area of science. 'Nano' is a prefix like 'milli' or 'mega'. While 'milli' means 'one-thousandth', 'nano' means 'one-thousand-millionth' – so nanoscience is the science of really tiny things.

What is nanoscience?

Our increasing understanding of science through the 20th century means that we now know that materials behave very differently at a very tiny scale. When we arrange atoms and molecules very carefully at this tiny scale, their properties can be truly remarkable.

Nanoscience at work

Glass can be coated with titanium oxide nanoparticles. Sunshine triggers a chemical reaction that breaks down dirt which then lands on the window. When it rains the water spreads evenly over the surface of the glass washing off the dirt.

Socks that are made from a fabric which contains silver nanoparticles never smell!

A type of lizard called a gecko can hang upside down from a sheet of glass. That's because the hairs on its feet are so tiny they can use the forces that hold molecules together. Scientists can make sticky tape lined with tiny nano-hairs that work in the same way.

Using nanoscience, health workers may soon be able to test a single drop of blood on a tiny piece of plastic no bigger than a ten pence piece. The tiny nanolab would replace individual tests for infectious diseases such as malaria and HIV/AIDS. On a larger scale these tests are both time-consuming and costly.

Nanoscience can do some pretty amazing things – these toy eyes are being moved using a tiny current from an electric battery

But some nanoscience is pure science fiction – tiny subs that travel through your blood to zap cancer cells with a laser; self-reproducing nanobots that escape and cover the Earth in 'grey goo' – only in airport novels!

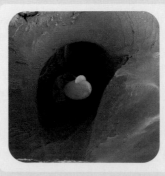

The science of tiny things – what should we do?

STUN
STAR IN SCANDAL SHOCK
We Find Out What They Don't Want You To Know... And WE TELL YOU!

IT'S ALL GOING GREY GOO...!

Boffins working on nanorobots reckon that there's a real danger that one day they will learn to reproduce.

When that happens, if the tiny creatures escape from the lab they may devour everything in sight, covering the world in grey goo…

A leading specialist in nanotechnology has warned that any

THE END OF THE LINE FOR DOCTORS?
R.I.P. G.Ps?!?

It could be the end of the line for your family doctor if nanotechnology carries on developing at this rate.

One day it may be possible to inject tiny robots into your blood. They'll work out what's wrong with you, send a message to a control centre outside your body and call for reinforcements to deal with what's wrong!

2010 WARRIORS

THE US Army is developing nanotech suits – thin uniforms which are flexible and tough enough to withstand bullets and blasts.

The uniforms would have GPS guidance systems and live satellite feeds of the battlefield piped directly into the soldier's brain. There is also a built-in air conditioning system to keep the body temperature normal. Inside the suit a full range of bio-sensors will send medical data back to a medical team.

Yesterday, a spokesman for the Pentag

CUTTING EDGE ENVIRONMENTAL NEWS EVERY WEEK!

NANOTECHNOLOGY GIVES CLEAN WATER

One-sixth of the world's population has no access to clean, safe water, and two million children die each year from water-related diseases. But nanoscience may come to the rescue.

Nano-membranes are portable and easily-cleaned systems that purify, detoxify and desalinate water far better than ordinary filters. Not only that – they are cheap too!

ACTIVITY

Whenever we are faced with a possible development in science there are two possible questions – what **can** we do? and what **should** we do?

Look at the ideas on the previous page and the four headlines on this page. Ask yourself these two questions about **one** of the headlines – and present your answers to your group.

SUMMARY QUESTIONS

1 Match the sentence halves together:

a) Ionic compounds have ……	A …… conduct electricity when molten or in solution.
b) Ionic compounds ……	B …… held together by strong electrostatic forces.
c) The oppositely charged ions in an ionic compound are ……	C …… a giant lattice of ions.
d) Ionic compounds are made of ……	D …… high melting points.

2 A certain ionic compound melts at exactly 800°C. Suggest how this compound could be used in a device to activate a warning light and buzzer when the temperature in a chemical reactor rises above 800°C.

3 The table contains data about some different substances:

Substance	Melting point (°C)	Boiling point (°C)	Electrical conductor
nickel	1455	2730	good
carbon dioxide	–	−78	poor
aluminium oxide	2072	2980	solid – poor liquid – good
copper	1083	2567	good
sodium bromide	747	1390	solid – poor liquid – good
silicon dioxide	1610	2230	poor
hydrogen chloride	−115	−85	poor
graphite	3652	4827	good

a) Make a table with the following headings: Giant covalent, Giant ionic, Molecular, Giant metallic. Now write the name of each substance above in the correct column.

b) One of these substances behaves in a slightly different way than its structure suggests – why?

4 'Both graphite and metals can conduct electricity – but graphite is soft while metals are not.' Use your knowledge of the different structures of graphite and metals to explain this statement. [Higher]

EXAM-STYLE QUESTIONS

1 The table contains information about some substances. Complete the missing information (a) to (g).

Melting point (°C)	Boiling point (°C)	Electrical conductivity when solid	Electrical conductivity when molten	Solubility in water	Type of bonding	Type of structure
1660	3287	(a)	good	insoluble	metallic	giant
−101	−35	poor	(b)	soluble	covalent	(c)
712	1418	poor	good	soluble	(d)	giant
−25	144	(e)	poor	insoluble	(f)	small molecules
1410	2355	poor	poor	insoluble	covalent	(g)

(7)

2 Quartz is a very hard mineral that is used as an abrasive. It is insoluble in water. It is a form of silica, SiO_2. It can form large, attractive crystals that are transparent and can be used for jewellery. It melts at 1610°C. It does not conduct electricity when solid or when molten. It is used in the form of sand in the building and glass-making industries.

(a) Give **three** pieces of evidence from the passage that tell you that quartz has a giant structure. (3)

(b) What type of bonding is in quartz? Explain your answer. (2)

3 Copper can be used to make electrical wires, water pipes, and cooking pans.

(a) Suggest **three** reasons why copper is used to make cooking pans. (3)

(b) Which **two** properties of copper depend on the ability of delocalised electrons to flow through the metal? (2)

(c) Explain what happens to the atoms in the metal when a piece of copper is pulled into a wire. (2)
[Higher]

4 Nanotechnology promises to revolutionise our world. Nanoparticles and new devices are being rapidly developed but production is still on a very small scale. The properties of nanoparticles that make them useful can cause problems if they are made in large quantities. These include explosions because of spontaneous combustion on contact with air.

(a) What are nanoparticles? (2)

(b) Suggest **two** reasons why nanotechnology is being developed rapidly. (2)

(c) Why are nanoparticles more likely to catch fire when exposed to air compared with normal materials? (2)

5 Piezoceramics are smart materials that can be made to vibrate by passing an electric current through them. They can be made small enough to work inside mobile phones.

(a) Suggest a possible economic advantage of piezoceramics. (1)

(b) Suggest an environmental advantage of piezoceramics. (1)

Some smart materials can only be seen at higher temperatures. They can be used in the manufacture of clothing.

(c) Suggest how this feature could be useful. (1)

6 A molecule of pentane can be represented as shown:

```
      H   H   H   H   H
      |   |   |   |   |
  H — C — C — C — C — C — H
      |   |   |   |   |
      H   H   H   H   H
```

(a) What do the letters C and H represent? (1)

(b) What do the lines between each C and H represent? (2)

(c) Explain why liquid pentane does not conduct electricity. (2)

(d) Pentane boils at 36°C. Explain what happens to the molecules of pentane when liquid pentane boils and becomes a gas. (2)

[Higher]

HOW SCIENCE WORKS QUESTIONS

A circuit was set up to test the conductivity of different solutions.

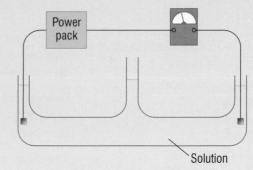

Two acids were tested to see how good they were at conducting electricity.
The results were set out in a graph:

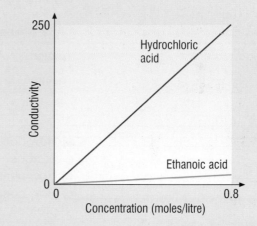

a) Describe the pattern shown by the graph for the hydrochloric acid. (1)

b) Could the relationship between the concentration of the hydrochloric acid and the conductivity be described as directly proportional? Explain your answer. (2)

c) i) What evidence is there, on the graph, that a solution of ethanoic acid does conduct electricity? (1)

   ii) What evidence is there that it does not conduct electricity as well as the hydrochloric acid conducts? (1)

d) What conclusion can you make about the conductivity of acids in general? (1)

e) A range of different acids were then tested. Their conductivity was measured at 0.4 moles/dm³. How would this data be presented? (1)

f) Does this evidence prove that acids separate into positive and negative ions? Explain your answer. (1)

C2 3.1

Mass numbers

LEARNING OBJECTIVES

1 What are the relative masses of protons, neutrons and electrons?
2 What is an atom's mass number?
3 What are isotopes?

As we saw earlier on, an atom consists of a nucleus containing positively charged protons, together with neutrons which have no charge. The negatively charged electrons are arranged in energy levels (shells) around the nucleus.

Every atom has the same number of electrons orbiting its nucleus as it has protons in its nucleus. The number of protons that an atom has is its **atomic number**.

The mass of a proton and a neutron are the same. Another way of putting this is to say that the *relative mass* of a neutron compared with a proton is 1. Electrons are far, far smaller than protons and neutrons – their mass is negligible. Because of this, the mass of an atom is concentrated in its nucleus. You can ignore the tiny mass of the electrons when it comes to thinking about the mass of an atom!

Type of sub-atomic particle	Relative mass
proton	1
neutron	1
electron	negligible (very small)

a) How does the number of electrons in an atom compare to the number of protons?
b) How does the mass of a proton compare to the mass of a neutron?
c) How does the mass of an electron compare to the mass of a neutron or proton?

Mass number

Almost all of the mass of an atom is found in the nucleus, because the mass of the electrons is so tiny. We call the total number of protons and neutrons in an atom its **mass number**.

When we want to show the atomic number and mass number of an atom we do it like this:

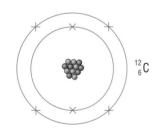

● Proton   Number of protons gives atomic number

● Neutron   Number of protons plus number of neutrons gives mass number

Figure 1 Chemists use the atomic number and the mass number of an element in many ways

Mass number $\quad ^{12}_{\ 6}C$ (carbon) $\quad ^{23}_{11}Na$ (sodium)
Atomic number

We can work out the number of neutrons in the nucleus of an atom by subtracting its atomic number from its mass number. The difference is the number of neutrons:

mass number − atomic number = number of neutrons

For the two examples here, carbon has 6 protons and a mass number of 12, so the number of neutrons is $(12 - 6) = 6$.

Sodium, on the other hand, has an atomic number of 11 but the mass number is 23, so $(23 - 11) = 12$. In this sodium atom there are 11 protons and 12 neutrons.

d) How do we calculate the number of neutrons in an atom?

Isotopes

Atoms of the same element always have the same number of protons, but they do not always have the same number of neutrons.

We give the name **isotopes** to atoms of the same element which have different numbers of neutrons.

For example, carbon has two common isotopes, $^{12}_{6}C$ (carbon-12) and $^{14}_{6}C$ (carbon-14). The carbon-12 isotope has 6 protons and 6 neutrons in the nucleus. The carbon-14 isotope has 6 protons and 8 neutrons.

Sometimes the extra neutrons in the nucleus make it unstable so that it is radioactive. However, not all isotopes are radioactive – they are simply atoms of the same substance with a different mass.

e) What are isotopes?

Different isotopes of the same element have different *physical* properties. For example, they have a different mass and they may be radioactive. However, they always have the same *chemical* properties.

For example, hydrogen has three isotopes: hydrogen, deuterium and tritium. (See Figure 2.) They each have a different mass and tritium is radioactive but they can all react with oxygen to make water.

f) Which isotope of hydrogen is heaviest?

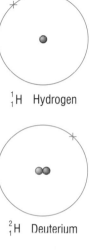

$^{1}_{1}H$ Hydrogen

$^{2}_{1}H$ Deuterium

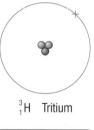

Figure 2 The isotopes of hydrogen – they have similar chemical properties but different physical properties

$^{3}_{1}H$ Tritium

SUMMARY QUESTIONS

1 Copy and complete using the words below:

   electrons   isotopes   mass   one

   The relative mass of a neutron compared to a proton is …… . Compared to protons and neutrons …… have almost no mass. The total number of protons and neutrons in an atom is called its …… number. Atoms of an element which have different numbers of neutrons are called …… .

2 State how many protons there would be in the nucleus of each of the following elements:

   a) $^{7}_{3}Li$,   b) $^{15}_{7}N$,   c) $^{22}_{10}Ne$,   d) $^{33}_{16}S$,   e) $^{79}_{35}Br$.

3 State how many neutrons each atom in question 2 has.

4 a) How do the physical properties of isotopes of the same element vary?
  b) Why do isotopes of the same element have identical chemical properties?

KEY POINTS

1 The relative mass of protons and neutrons is 1.
2 The mass number of an atom tells you the total number of protons and neutrons in its nucleus.
3 Isotopes are atoms of the same element with different numbers of neutrons.

Masses of atoms and moles

LEARNING OBJECTIVES

1 How can we compare the mass of atoms? [Higher]
2 How can we calculate the mass of compounds from the elements they are made from?

Chemical equations show you how many atoms of the reactants we need to make the products. But when we actually carry out a chemical reaction we need to know what amounts to use in grams or cm³. You might think that a chemical equation would also tell you this.

For example, does the equation:

$$Mg + 2HCl \rightarrow MgCl_2 + H_2$$

mean that we need twice as many grams of hydrochloric acid as magnesium to make magnesium chloride?

Unfortunately it isn't that simple. The equation tells us that we need twice as many hydrogen and chlorine atoms as magnesium atoms – but this doesn't mean that the mass of hydrochloric acid will be twice the mass of magnesium. This is because atoms of different elements have different masses.

To turn equations into something that we can actually use in the lab or factory we need to know a bit more about the mass of atoms.

a) Why don't chemical equations tell us how much of each reactant to use in a chemical reaction?

Relative atomic masses

The mass of a single atom is so tiny that it would be impossible to use it in calculations. To make the whole thing manageable we use a much simpler way of thinking about the masses of atoms. Instead of working with the *real* masses of atoms we just focus on the *relative* masses of atoms of different elements. We call these **relative atomic masses (A_r)**.

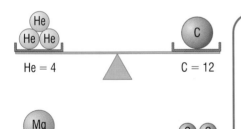

Figure 1 The A_r of carbon is 12. Compared with this, the A_r of helium is 4 and the A_r of magnesium is 24

Relative atomic mass	Relative ionic mass
Na 23	Na⁺ 23
O 16	O²⁻ 16
Mg 24	Mg²⁺ 24

We use an atom of carbon ($^{12}_6C$) as a standard atom. We give this a 'mass' of 12 units, because it has 6 protons and 6 neutrons. We then compare all of the masses of the atoms of all the other elements to this standard carbon atom.

The mass of an atom found by comparing it with the $^{12}_6C$ atom is called its relative atomic mass (A_r).

The relative atomic mass of an element is usually the same as, or very similar to, the mass number of that element. The A_r takes into account any isotopes of the element. The relative atomic mass is the average mass of the isotopes of the element in the proportions in which they are usually found (compared with the standard carbon atom).

When atoms change into ions they either lose or gain electrons. However, for all practical purposes the mass of electrons isn't worth bothering about. So the 'relative ionic mass' of an ion is exactly the same as the relative atomic mass of that element.

b) What do we call the mass of an atom compared with the mass of an atom of carbon-12?

Relative formula masses

We can use the A_r of the various elements to work out the **relative formula mass (M_r)** of chemical compounds. This is true whether the compounds are made up of molecules or collections of ions. A simple example is a substance like sodium chloride. We know that the A_r of sodium is 23 and the A_r of chlorine is 35.5. So the relative formula mass of sodium chloride (NaCl) is:

$$23 + 35.5 = 58.5$$
$$A_r\ \text{Na} \quad A_r\ \text{Cl} \quad M_r\ \text{NaCl}$$

Another example is water. Water is made up of hydrogen and oxygen. The A_r of hydrogen is 1, and the A_r of oxygen is 16. Water has the formula H_2O, containing two hydrogen atoms for every one oxygen, so the M_r is:

$$(1 \times 2) + 16 = 18$$
$$A_r\ \text{H} \times 2 \quad A_r\ \text{O} \quad M_r\ \text{H}_2\text{O}$$

c) What is the relative formula mass of hydrogen sulfide, H_2S?
   (A_r values: H = 1, S = 32)

Moles

Saying or writing 'relative atomic mass in grams' or 'relative formula mass in grams' is rather clumsy. So chemists have a shorthand word for it – **mole**.

They say that the relative atomic mass in grams of carbon (i.e. 12 g of carbon) is a mole of carbon atoms. One mole is simply the relative atomic mass or relative formula mass of any substance expressed in grams. A mole of any substance always contains the same number of particles.

Figure 2 We know how many actual atoms or molecules a mole contains, thanks to an Italian count born in the 18th century, Amedeo Avogadro. He worked out that a mole of any element or compound contains 6.02×10^{23} atoms, ions or molecules. That's 602 000 000 000 000 000 000 000! This is called **Avogadro's number**.

We can use the same approach with relatively complicated molecules like sulfuric acid, H_2SO_4. Hydrogen has a A_r of 1, the A_r of sulfur is 32 and the A_r of oxygen 16. This means that the M_r of sulfuric acid is:

$$(1 \times 2) + 32 + (16 \times 4) =$$
$$2 + 32 + 64 = 98$$

GET IT RIGHT!

You don't have to remember Avogadro's number! But practise calculating the mass of one mole of different substances from relative atomic masses that you are given.

SUMMARY QUESTIONS

1   Copy and complete using the words below:

   atomic   carbon-12   elements   formula   number

   We measure the masses of atoms by comparing them to the mass of one atom of ……. . The relative …… mass of an element is usually almost the same as its mass ……. . We calculate the relative …… mass of a compound from the relative atomic masses of the …… in it.

2   The equation for the reaction of magnesium and fluorine is:

   $$Mg + F_2 \rightarrow MgF_2$$

   a) How many moles of fluorine molecules react with one mole of magnesium atoms?
   b) What is the relative formula mass of MgF_2? (A_r values: Mg = 24, F = 19)

3   The relative atomic mass of oxygen is 16, and that of magnesium is 24. How many times heavier is a magnesium atom than an oxygen atom?

KEY POINTS

1   We compare the masses of atoms by measuring them relative to atoms of carbon-12. [Higher]
2   We work out the relative formula mass of a compound from the relative atomic masses of the elements in it.
3   One mole of any substance always contains the same number of particles.

C2 3.3   Percentages and formulae

LEARNING OBJECTIVES

1 How can we calculate the percentage mass of each element in a compound from its formula?
2 How can we calculate the formula of a compound from its percentage composition? [Higher]

Figure 1 A tiny difference in the amount of iron in the ore might not seem very much, but when millions of tonnes of iron ore are extracted and processed each year, it all adds up!

To calculate the percentage of an element in a compound:

- Write down the formula of the compound.
- Using the relative atomic masses from your data sheet work out the relative formula mass of the compound. Write down the mass of each element making up the compound as you work it out.
- Write the mass of the element you are investigating as a fraction of the total M_r.
- Find the percentage by multiplying your fraction by 100.

We can use the formula mass of a compound to calculate the percentage mass of each element in it. Calculations like this are not just done in GCSE chemistry books! In life outside the school laboratory, geologists and mining companies base their decisions about whether to exploit mineral finds on calculations like these.

Working out the amount of an element in a compound

We can use the relative atomic mass (A_r) of elements and the relative formula mass (M_r) of compounds to help us work out the percentage of an element in a compound.

Worked example (1)

What percentage mass of white magnesium oxide is actually magnesium, and how much is oxygen?

Solution

The first thing we need is the formula of magnesium oxide, MgO.
The A_r of magnesium is 24, while the A_r of oxygen is 16.

Adding these together gives us a M_r of 40 i.e. (24 + 16).

So from 40 g of magnesium oxide, 24 g is actually magnesium:

$$\frac{\text{mass of magnesium}}{\text{total mass of compound}} = \frac{24}{40}$$

so the percentage of magnesium in the compound is:

$$\frac{24}{40} \times 100\% = 60\%$$

Worked example (2)

A white powder is found at the scene of a crime. It could be strychnine, a deadly poison with the formula $C_{21}H_{22}N_2O_2$ – but is it?!

When a chemist analyses the powder, 83% of its mass is carbon. What is the percentage mass of carbon in strychnine, and is this the same?

Solution

The formula mass (M_r) of strychnine is:

$$(12 \times 21) + (1 \times 22) + (14 \times 2) + (16 \times 2) = 252 + 22 + 28 + 32 = 334$$

The percentage mass of carbon in strychnine is therefore:

$$\frac{252}{334} \times 100 = 75.4\%$$

This is **not** the same as the percentage mass of carbon in the white powder – so the white powder is not strychnine.

---

a) What is the percentage mass of hydrogen in methane, CH_4?
   (A_r values: C = 12, H = 1)

HIGHER

Working out the formula of a compound from its percentage composition

We can also do this backwards! If we know the percentage composition of a compound we can work out the ratio of the numbers of atoms in the compound. We call this its **empirical formula**. It tells us the simplest whole number ratio of elements in a compound.

This is sometimes the same as the actual number of atoms in one molecule (which we call the **molecular formula**) – but not always. For example, the empirical formula of water is H_2O, which is also its molecular formula. However, hydrogen peroxide has the empirical formula HO, but its molecular formula is H_2O_2.

Figure 2 Chemical analysis of substances found at the scene of a crime may help to bring a murderer to justice – or free an innocent suspect

Worked example

If 9 g of aluminium react with 35.5 g of chlorine, what is the empirical formula of the compound formed?

Solution

We can work out the ratio of the number of atoms by dividing the mass of each element by its relative atomic mass:

$$\text{For aluminium: } \frac{9}{27}\,g = \tfrac{1}{3} \text{ mole of aluminium atoms}$$

$$\text{For chlorine: } \frac{35.5}{35.5}\,g = 1 \text{ mole of chlorine atoms}$$

So this tells us that one mole of chlorine atoms combines with $\tfrac{1}{3}$ mole of aluminium atoms.

This means that the simplest whole number ratio is 3 (Cl) : 1 (Al).
In other words 1 aluminium atom combines with 3 chlorine atoms. So the empirical formula is $AlCl_3$.

b) A compound contains 16 g of sulfur and 24 g of oxygen. What is its empirical formula? (A_r values: S = 32, O = 16)

Given the percentage composition and asked to find the empirical formula, just assume you have 100 g of the compound. Then do a calculation as shown above to find the simplest ratio of elements.

GET IT RIGHT!

Make sure that you can do these calculations from formula to percentage mass and the other way round.

How to work out the formula from reacting masses:

- Begin with the number of grams of the elements that combine.
- Change the number of grams to the moles of atoms by dividing the number of grams by the A_r. This tells you how many moles of the different elements combine.
- Use this to tell you the simplest ratio of atoms of the different elements combined in the compound.
- This gives you the empirical formula of the compound.

SUMMARY QUESTIONS

1 Copy and complete using the words below:

   compound   dividing   hundred   relative formula mass

   The percentage of an element in a …… is calculated by …… the mass of the element in the compound by the …… …… …… of the compound and then multiplying the result by one …… .

2 Ammonium nitrate (NH_4NO_3) is used as a fertiliser. What is the percentage mass of nitrogen in it? (A_r values: H = 1, N = 14, O = 16)

3 22.55% of the mass of a sample of phosphorus chloride is phosphorus. What is the formula of phosphorus chloride? (A_r values: P = 31, Cl = 35.5)
   [Higher]

KEY POINT

1 The relative atomic masses of the elements in a compound can be used to work out its percentage composition.
2 We can calculate empirical formulae given the masses or percentage composition of elements present. [Higher]

C2 3.4   Equations and calculations

LEARNING OBJECTIVES

1 What do chemical equations tell us about chemical reactions?
2 How do we use equations to calculate masses of reactants and products?

Chemical equations can be very useful when we want to know how much of each substance is involved in a chemical reaction. But to do this, we must be sure that the equation is balanced.

To see how we do this, think about what happens when hydrogen molecules (H_2) react with oxygen molecules (O_2), making water molecules (H_2O):

$$H_2 + O_2 \rightarrow H_2O \text{ (not balanced)}$$

This equation shows the reactants and the product – but it is not balanced. There are 2 oxygen atoms on the left-hand side and only 1 oxygen atom on the right-hand side. To balance the equation there need to be 2 water molecules on the right-hand side:

$$H_2 + O_2 \rightarrow 2H_2O \text{ (still not balanced)}$$

This balances the number of oxygen atoms on each side of the equation – but now there are 4 hydrogen atoms on the right-hand side and only 2 on the left-hand side. So there need to be 2 hydrogen molecules on the left-hand side:

$$2H_2 + O_2 \rightarrow 2H_2O \text{ (balanced!)}$$

This balanced equation tells us that '2 hydrogen molecules react with one oxygen molecule to make 2 water molecules'. But remember that 1 mole of any substance always contains the same number of particles. So our balanced equation also tells us that '2 moles of hydrogen molecules react with one mole of oxygen molecules to make two moles of water molecules'.

a) What must we do to a chemical equation before we use it to work out how much of each chemical is needed or made?
b) '$2H_2$' has two meanings – what are they?

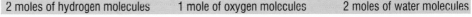

2 hydrogen molecules	1 oxygen molecule	2 water molecules
$2H_2$	$+$ O_2 $\longrightarrow$	$2H_2O$
2 moles of hydrogen molecules	1 mole of oxygen molecules	2 moles of water molecules

This is really useful, because we can use it to work out what mass of hydrogen and oxygen we need, and how much water is made.

To do this, we need to know that the A_r for hydrogen is 1 and the A_r for oxygen is 16:

A_r of hydrogen = 1 ...... so mass of 1 mole of $H_2 = 2 \times 1 = 2\,g$
A_r of oxygen = 16 ...... so mass of 1 mole of $O_2 = 2 \times 16 = 32\,g$
M_r of water = (16 + 2) = 18 ...... so mass of 1 mole of water = 18 g

Our balanced equation tells us that 2 moles of hydrogen react with one mole of oxygen to give 2 moles of water. So turning this into masses we get:

2 moles of hydrogen = $2 \times 2\,g = 4\,g$
1 mole of oxygen   = $1 \times 32\,g = 32\,g$
2 moles of water   = $2 \times 18\,g = 36\,g$

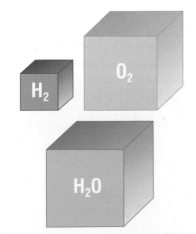

Figure 1 When 4 g of hydrogen react with 32 g of oxygen we get 36 g of water

Calculations

These kind of calculations are important when we want to know how much of two chemicals to react together. For example, a chemical called sodium hydroxide reacts with chlorine gas to make bleach.

Here is the equation for the reaction:

$$2\,NaOH + Cl_2 \rightarrow NaOCl + NaCl + H_2O$$

sodium   chlorine   bleach   salt   water
hydroxide

This reaction happens when chlorine gas is bubbled through a solution of sodium hydroxide dissolved in water.

If we have a solution containing 100 g of sodium hydroxide, how much chlorine gas should we pass through the solution to make bleach? Too much, and some chlorine will be wasted, too little and not all of the sodium hydroxide will react.

	So mass of 1 mole of	
	NaOH	**Cl_2**
A_r of hydrogen = 1		
A_r of oxygen = 16	= 23 + 16 + 1 = 40	= 35.5 × 2 = 71
A_r of sodium = 23		
A_r of chlorine = 35.5		

The table shows that 1 mole of sodium hydroxide has a mass of 40 g.

So 100 g of sodium hydroxide is $\dfrac{100}{40} = 2.5$ moles.

The chemical equation for the reaction tells us that for every two moles of sodium hydroxide we need one mole of chlorine.

So we need $\dfrac{2.5}{2} = 1.25$ moles of chlorine.

The table shows that 1 mole of chlorine has a mass of 71 g.

So we will need 1.25 × 71 = **88.75 g** of chlorine to react with 100 g of sodium hydroxide.

DID YOU KNOW?

Mole is the English version of the German word *Mol* which is short for *Molekulargewicht*, the 'molecular weight.'

Figure 2 Bleach is used in some swimming pools to control and kill harmful bacteria. Getting the quantities right involves some careful calculation!

SUMMARY QUESTIONS

1 Copy and complete using the words below:

   balanced   equations   mole   mass   product

   Chemical ...... can tell us about the amount of substances in a reaction if they are .... To work out the mass of each substance in a reaction we need to know the mass of 1 ...... of it. We can then work out the ...... of each reactant needed, and the mass of ...... that will be produced.

2 Hydrogen peroxide, H_2O_2, decomposes to form water and oxygen gas. Write a balanced equation for this reaction.

3 Calcium reacts with oxygen like this:

$$2\,Ca + O_2 \rightarrow 2\,CaO$$

   What mass of oxygen will react exactly with 60 g of calcium? (A_r values: O = 16, Ca = 40)

KEY POINTS

1 Chemical equations tell us the number of moles of substances in the chemical reaction.

2 We can use chemical equations to calculate the masses of reactants and products in a chemical reaction from the masses of one mole of each of the substances involved in the reaction.

C2 3.5   Making as much as we want

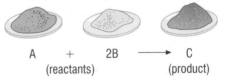

LEARNING OBJECTIVES

1 What do we mean by the yield of a chemical reaction and what factors affect it?
2 How do we calculate the yield of a chemical reaction? [Higher]
3 What is atom economy and why is it important?
4 How do we calculate atom economy? [Higher]

Many of the substances that we use every day have to be made from other chemicals, using complex chemical reactions. Food colourings, flavourings and preservatives, the ink in your pen or computer printer, the artificial fibres in your clothes – all of these are made using chemical reactions.

One simple kind of reaction for making a new substance is when we make a new chemical from two others, like this:

$$A + 2B \longrightarrow C$$

(reactants)   (product)

If we need 1000 kg of C it seems quite simple for us to work out how much A and B we need to make it. As we saw earlier in this chapter, all we need to know is the relative formula masses of A, B and C.

a) How many moles of B are needed to react with each mole of A in this reaction?

b) How many moles of C will this make?

If we carry out our reaction, it is very unlikely that we will get as much of C as we think. This is because our calculations assumed that **all** of A and B would be turned into C. We call the amount of product that a chemical reaction produces its **yield**.

Calculating percentage yield

Rather than talking about the yield of a chemical reaction in grams, kilograms or tonnes it is much more useful to talk about its **percentage yield**. This compares the amount of product that the reaction *really* produces with the maximum amount that it could *possibly* produce:

$$\text{percentage yield} = \frac{\text{amount of product produced}}{\text{maximum amount of product possible}} \times 100\%$$

Worked example

Using known masses of A and B, it was calculated that the chemical reaction above could produce 2.5 g of product, C. When the reaction is carried out, only 1.5 g of C is produced.

What is the percentage yield of this reaction?

Solution

$$\text{Percentage yield} = \frac{\text{amount of product produced}}{\text{maximum amount of product possible}} \times 100\%$$

$$= \frac{1.5}{2.5} \times 100\%$$

$$= 60\%$$

The percentage yield is **60%**.

c) How is percentage yield calculated?

Very few chemical reactions have a yield of 100% because:

- The reaction may be reversible (so as products form they react to form the reactants again).
- Some reactants may react to give unexpected products.
- Some of the product may be left behind in the apparatus.
- The reactants may not be completely pure.
- Some chemical reactions produce more than one product, and it may be difficult to separate the product that we want from the reaction mixture.

Atom economy

Chemical companies use chemical reactions to make products which they sell. So it is very important to use chemical reactions that produce as much product as possible. In other words, it is better for them to use chemical reactions with high yields.

Making as much product as possible means making less waste. It means that as much product as possible is being made from the reactants. This is good news for the company's finances, and good news for the environment too.

The amount of the starting materials that end up as useful products is called the **atom economy**. So the aim is to achieve maximum atom economy.

> We can calculate percentage atom economy using this equation:
>
> $$\text{percentage atom economy} = \frac{\text{relative formula mass of useful product}}{\text{relative formula mass of all products}} \times 100$$
>
> **HIGHER**
>
> > **Worked example**
> > Ethanol (C_2H_5OH) can be converted into ethene (C_2H_4) which can be used to make poly(ethene).
> >
> > **Solution**
> >
> > $$C_2H_5OH \longrightarrow C_2H_4 \quad + \quad H_2O$$
> >
> > M_r values: $\quad (12 \times 2) + (1 \times 4) \quad (1 \times 2) + (16 \times 1)$
> > $\qquad\qquad\qquad = 28 \qquad\qquad\qquad = 18$
> >
> > $$\text{percentage atom economy} = \frac{28}{(28 + 18)} \times 100 = 61\%$$
>
> To conserve the Earth's resources, as well as reduce pollution and waste, industry tries to maximise both atom economy and percentage yield.

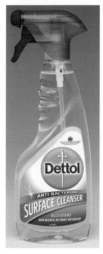
Figure 1 When you make and sell large quantities of chemicals, it's important to know the yield of the reactions you are using

KEY POINTS

1 The yield of a chemical reaction describes how much product is made.
2 The percentage yield of a chemical reaction tells us how much product is made compared with the maximum amount that could be made (100%). [Higher]
3 Factors affecting the yield of a chemical reaction include product being left behind in the apparatus and difficulty separating the products from the reaction mixture.
4 It is important to maximise atom economy to conserve resources and reduce pollution.

SUMMARY QUESTIONS

1 Copy and complete using the words below:

   high   maximum   percentage   product   waste   yield

   The amount of ...... made in a chemical reaction is called its .... The ...... yield tells us the amount of product that is made compared to the ...... amount that could be made. Reactions with ...... yields are important because they make less ....

2 A reaction produces a product which has a relative formula mass of 80. The total of the relative formula masses of all the products is 120. What is the percentage atom economy of this reaction? [Higher]

3 A reaction that could produce 200 g of product produces only 140 g. What is its percentage yield? [Higher]

4 If the percentage yield for a reaction is 100%, 60 g of reactant A would make 80 g of product C. How much of reactant A is needed to make 80 g of product C if the percentage yield of the reaction is only 75%? [Higher]

C2 3.6

Reversible reactions

1 What is a reversible reaction?
2 How can we change the amount of product in a reversible reaction? [Higher]

In all of the chemical reactions that we have looked at so far the reactants have reacted together to form products. We show this by using an arrow pointing *from* the reactants *to* the products, like this:

$$A + B \rightarrow C + D$$

But in some chemical reactions the products can react together to produce the original reactants again. We call this a **reversible reaction**.

Because a reversible reaction can go in both directions we use two arrows in the equation, one going in the forwards direction and one backwards instead of the usual single arrow:

$$A + B \rightleftharpoons C + D \quad (\rightleftharpoons \text{ is equilibrium sign})$$

a) What does a single arrow in a chemical equation mean?
b) What does a double arrow in a chemical equation mean?

HIGHER

So what happens when we start with just reactants in a reversible reaction?

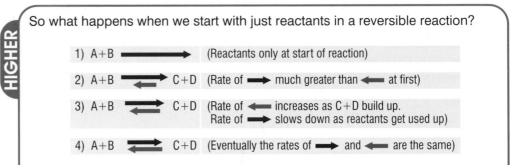

1) A+B ⟶ (Reactants only at start of reaction)

2) A+B ⇄ C+D (Rate of ⟶ much greater than ⟵ at first)

3) A+B ⇄ C+D (Rate of ⟵ increases as C+D build up. Rate of ⟶ slows down as reactants get used up)

4) A+B ⇄ C+D (Eventually the rates of ⟶ and ⟵ are the same)

In a system that is **closed**, no reactants or products can get in or out. In a closed system, as more and more products are made in a reversible reaction the rate at which these get converted back into reactants increases. As the rate of the backward (reverse) reaction increases, the rate of the forward reaction decreases until both reactions are going at the same rate.

When this happens the reactants are making products at the same rate as the products are making reactants again – so overall there is no change in the amount of products and reactants. We say that the reaction is at the point of **equilibrium**. At equilibrium the rate of the forward reaction equals the rate of the reverse reaction.

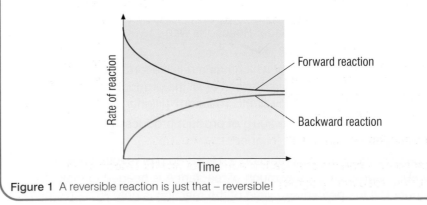

Figure 1 A reversible reaction is just that – reversible!

HIGHER

c) How does the rate of the backward reaction compare to the rate of the forward reaction at equilibrium?

One example of a reversible reaction is the reaction between iodine monochloride (ICl) and chlorine gas. Iodine monochloride is a brown liquid, while chlorine is a green gas. We can react these substances together to make yellow crystals of iodine trichloride (ICl$_3$).

When there is plenty of chlorine gas the forward reaction makes iodine trichloride crystals which are quite stable. But if we lower the concentration of chlorine gas the backward (reverse) reaction turns iodine trichloride back to iodine monochloride and chlorine.

Figure 2 The situation at equilibrium is just like running up an escalator which is going down – if you run *up* as fast as the escalator goes *down*, you will get *nowhere*!

DID YOU KNOW?

Many of the chemical reactions that take place in your body are reversible. The rate of the reaction in each direction is controlled by special chemicals called **enzymes**.

Figure 3 This equilibrium can be changed by adding or removing chlorine from the reaction

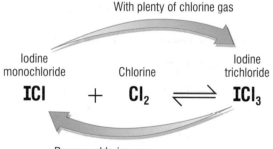

With plenty of chlorine gas

| Iodine monochloride | | Chlorine | | Iodine trichloride |
| ICl | + | Cl$_2$ | $\rightleftharpoons$ | ICl$_3$ |

Remove chlorine gas

We can change the relative proportions of the reactants and products in a reaction mixture by changing the reaction conditions. This is very important, because if we want to collect the products of a reaction we need as much product as possible in the reacting mixture.

SUMMARY QUESTIONS

1 Copy and complete using the words below:

> **amount   conditions   equilibrium   forward   products**
> **rate   reactants   reverse   reversible**

In some chemical reactions the …… can react to form the reactants again. We call this a …… reaction. At …… in a closed system, the …… of the …… reaction is the same as the rate of the …… reaction. If we change the reaction ……, this can affect the …… of the …… and products in the mixture.   [Higher]

2 What does the $\rightleftharpoons$ sign mean in a chemical equation?

3 In general how can we change the amount of product made in an equilibrium reaction? [Higher]

KEY POINTS

1 In a reversible reaction the products of the reaction can react to make the original reactants.

2 In a closed system the rate of the forward and backward (reverse) reactions are equal at equilibrium. [Higher]

3 Changing the reaction conditions can change the amounts of products and reactants in a reaction mixture. [Higher]

C2 3.7

Making ammonia – the Haber process

We need plants – for food, and as a way of providing the oxygen that we breathe. Plants need nitrogen to grow, and although this gas makes up about 80% of the air around us, plants cannot use it because it is very unreactive.

Instead, plants absorb soluble nitrates from the soil through their roots. When we harvest plants these nitrates are lost – so we need to replace them. Nowadays we usually do this by adding nitrate fertilisers to the soil. We make these fertilisers using a process invented nearly 100 years ago by a young German chemist called Fritz Haber.

a) Why can't plants use the nitrogen from the air?
b) Where do plants get their nitrogen from?

Figure 1 Plants are surrounded by nitrogen in the air. They cannot use this nitrogen, and rely on soluble nitrates in the soil instead. We supply these by spreading fertiliser on the soil.

The Haber process

The Haber process provides us with a way of turning the nitrogen in the air into ammonia. We can use ammonia in many different ways. One of the most important of these is to make fertilisers.

The raw materials for making ammonia are:

• nitrogen from the air, and
• hydrogen which we get from natural gas (containing mainly methane, CH_4).

The nitrogen and hydrogen are purified and then passed over an iron catalyst at high temperatures (about 450°C) and pressures (about 200 atmospheres). The product of this chemical reaction is ammonia.

c) What are the two raw materials needed to make ammonia?

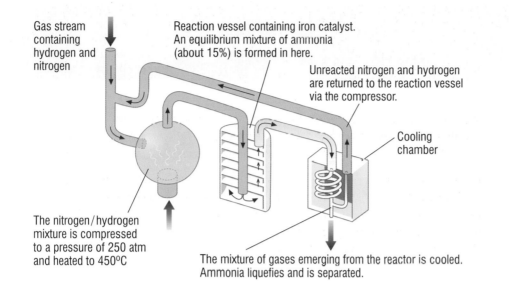

Gas stream containing hydrogen and nitrogen

Reaction vessel containing iron catalyst. An equilibrium mixture of ammonia (about 15%) is formed in here.

Unreacted nitrogen and hydrogen are returned to the reaction vessel via the compressor.

Cooling chamber

The nitrogen/hydrogen mixture is compressed to a pressure of 250 atm and heated to 450°C

The mixture of gases emerging from the reactor is cooled. Ammonia liquefies and is separated.

Figure 2 The Haber process

The reaction used in the Haber process is reversible, which means that the ammonia breaks down again into hydrogen and nitrogen. To reduce this, we have to remove the ammonia by cooling and liquefying it as soon as it is formed. We can then recycle any hydrogen and nitrogen that is left so that it has a chance to react again.

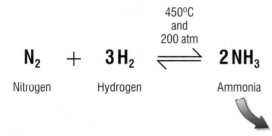

$$N_2 \ + \ 3H_2 \ \underset{}{\overset{450°C \ and \ 200 \ atm}{\rightleftharpoons}} \ 2NH_3$$

Nitrogen   Hydrogen   Ammonia

By removing the ammonia that forms we can reduce the rate of the backwards reaction. This helps to stop the ammonia that is formed from breaking down into nitrogen and hydrogen.

We carry out the Haber process in conditions that have been carefully chosen to give a reasonable yield of ammonia as quickly as possible. (See pages 178–9.)

d) How is ammonia removed from the reaction mixture?

e) How do we make sure the reactants are not wasted?

SUMMARY QUESTIONS

1 Copy and complete using the words below:

air   fertilisers   gas   450   hydrogen   iron   liquefying   nitrogen   removed   200

Ammonia is an important chemical used for making .... The raw materials are ...... from the ...... and ...... from natural .... These are reacted at about ...... °C and ...... atmospheres pressure using an ...... catalyst. Ammonia is ...... from the reaction mixture before it can break down into the reactants again by ...... the gas.

2 Draw a flow diagram to show how the Haber process is used to make ammonia.

KEY POINTS

1 Ammonia is an important chemical for making other chemicals, including fertilisers.
2 Ammonia is made from nitrogen and hydrogen in the Haber process.
3 We carry out the Haber process under conditions which are chosen to give a reasonable yield of ammonia as quickly as possible.
4 Any unused nitrogen and hydrogen are recycled in the Haber process.

Using reversible reactions – 'green' chemistry

- Using relative atomic masses we can work out the relative formula mass of compounds so that we know the mass of 1 mole

- Balanced chemical equations show us the number of moles of reactants and products in reactions

- We can make more products from a reversible reaction by removing them as they are formed

- The yield of a chemical reaction tells us how much product a reaction will REALLY give us compared with what it could POSSIBLY give us

- We use chemical equations to work out what mass of each reactant we need to make as much product as we want

- Some chemical reactions go both ways – they are REVERSIBLE

- Processes with a high percentage atom economy are more profitable and are better for sustainable development

- By carefully choosing chemical reactions and the conditions, we can make as much product from a chemical reaction as possible – with as little waste as possible

ACTIVITY

Look at the ideas on this page. Use them to design a poster about 'Crafty Chemists', showing how chemistry helps us to use chemical reactions which make as much of what we want as efficiently as possible.

Fritz Haber – a good life or a bad one?

Der Newswell

Donnerstag 14. Oktober 1920

Fritz Haber - German Patriot

Early on in the First World War both sides became bogged down in trench warfare. Fritz Haber focused on what he could do to bring about German victory. He thought that poison gas would penetrate the strongest defences, allowing Germany to win the war.

Poison gases were already available as unwanted by-products of chemical processes. Haber experimented with these gases to find those suitable to use on the battlefield. He focused on chlorine gas.

The Germans used chlorine for the first time on 22nd April 1915, against French and Algerian troops in Belgium. They released 200 tonnes of gas which rolled into the allied lines. The allied soldiers choked in agony, and slowly died. The gas cloud tinted everything a sickly green. Those who could escape the cloud fled in panic.

The soldier Wilfrid Owen wrote:

GAS! Gas! Quick, boys! – An ecstasy of fumbling,
Fitting the clumsy helmets just in time;
But someone still was yelling out and stumbling
And floundering like a man in fire or lime.
Dim, through the misty panes and thick green light
As under a green sea, I saw him drowning.
In all my dreams, before my helpless sight,
He plunges at me, guttering, choking, drowning.

Tragedy of Clara Haber

19th November 1920 1d   Palestine becomes British mandate - page 4

Clara Haber had graduated as the first woman from the University of Breslau in Germany in 1900. At this time many professors were still against female students. She married Fritz Haber in 1901. Clara could not continue her research work because she was a woman. Instead, she contributed to her husband's work, although her support was never mentioned. Because of her support Haber was promoted very quickly.

Clara was deeply opposed to warfare, and especially to the use of science in war. She called Haber's work a 'perversion of science'. Unable to stop him, Clara committed suicide in May 1915. Her death was never announced in the newspapers.

League of Nations holds first meeting in Geneva, Switzerland

ACTIVITY

A new laboratory has been set up to develop a new chemical to be used as a weapon.

Either: As a brilliant young chemist you have been invited to go and work in this laboratory. Write a letter or e-mail to a friend explaining why you are planning to accept this invitation.

Or: A friend has been invited to go and work in this laboratory. Write a letter or e-mail to this friend explaining why you think they should not go and work in this laboratory.

SUMMARY QUESTIONS

1 Match up the parts of the sentences:

a) Neutrons have a relative mass of	A ...... negligible mass compared to protons and neutrons.
b) Electrons have	B ...... 1 compared to protons.
c) Protons have a relative mass of	C ...... found in its nucleus.
d) Nearly all of an atom's mass is	D ...... 1 compared to neutrons.

2 Calculate the mass of 1 mole of each of the following compounds:

a) CO_2

b) B_2H_6

c) $CaCO_3$

d) K_2O

e) $KMnO_7$

(A_r values: C = 12, O = 16, B = 11, H = 1, Ca = 40, K = 39, Mn = 55)

3 How many moles of

a) Ag atoms are there in 54 g of silver,

b) P atoms are there in 3.1 g of phosphorus,

c) Fe atoms are there in 0.56 g of iron?

(A_r values: Ag = 108, P = 31, Fe = 56)

4 When aluminium reacts with bromine, 1.35 g of aluminium reacts with 12.0 g of bromine. What is the empirical formula of aluminium bromide?

(A_r values: Al = 27, Br = 80)   [Higher]

5 In a lime kiln, calcium carbonate is decomposed to calcium oxide:

$CaCO_3 \rightarrow CaO + CO_2$

Calculate the percentage atom economy for the process (assuming 100% conversion).

(A_r values: Ca = 40, O = 16, C = 12)   [Higher]

6 a) What is a reversible reaction?

b) How does a reversible reaction differ from an 'ordinary' reaction?

c) State the conditions chosen for the Haber process to convert nitrogen and hydrogen into ammonia.

EXAM-STYLE QUESTIONS

1 Hydrogen has three isotopes, ^1_1H, ^2_1H, and ^3_1H.

(a) What are isotopes?   (2)

(b) How many protons, neutrons and electrons are there in an atom of ^3_1H?   (3)

(c) Heavy water contains atoms of the isotope ^2_1H instead of ^1_1H. It has the formula $^2_1H_2O$. What is the mass of one mole of heavy water?   (2)

2 Tablets taken by people with iron deficiency anaemia contain 0.200 g of anhydrous iron(II) sulfate, $FeSO_4$.

(a) Calculate the relative formula mass of iron(II) sulfate, $FeSO_4$.   (2)

(b) Calculate the percentage of iron in iron(II) sulfate.   (2)

(c) Calculate the mass of iron in each tablet.   (2)

3 The equation for the main reaction to make ammonia is:

$$N_2 + 3H_2 \rightleftharpoons 2NH_3$$

(a) What does the symbol $\rightleftharpoons$ tell you about this reaction?   (1)

(b) The flow diagram shows the main stages in making ammonia.

(i) Name the two raw materials **A** and **B**.   (2)

(ii) What is the purpose of the iron in the reactor?   (1)

(iii) Why do the nitrogen and hydrogen not react completely.   (1)

(iv) How is wastage of unreacted nitrogen and hydrogen prevented?   (1)

4 The equation for the reaction of calcium carbonate with hydrochloric acid is:

$$CaCO_3 + 2HCl \rightarrow CaCl_2 + CO_2 + H_2O$$

(a) How many moles of hydrochloric acid react with one mole of calcium carbonate? (1)

(b) How many moles of calcium chloride are produced from one mole of calcium carbonate? (1)

(c) What is the mass of calcium chloride that can be made from one mole of calcium carbonate? (2)

(d) What is the mass of one mole of calcium carbonate? (2)

(e) A student reacted 10 g of calcium carbonate with hydrochloric acid and collected 7.4 g of calcium chloride. What was the percentage yield? (2)
[Higher]

5 Chromium can be obtained from chromium oxide, Cr_2O_3, by reduction with aluminium or carbon. For the first reaction, chromium is mixed with aluminium and ignited in a crucible. The reaction using carbon is done at high temperatures in a low-pressure furnace. The equations for the reactions are:

$$Cr_2O_3 + 2Al \rightarrow 2Cr + Al_2O_3$$
$$2Cr_2O_3 + 3C \rightarrow 4Cr + 3CO_2$$

(a) Calculate the maximum mass of chromium that can be obtained from one mole of chromium oxide. (2)

(b) Calculate the percentage atom economy for both reactions to show which reaction has the better atom economy. (4)

(c) Suggest one advantage and one disadvantage of using carbon to manufacture chromium. (2)
[Higher]

6 Ibuprofen is used as a pain killer throughout the world. You might know it as Nurofen or Ibuleve. The traditional way to manufacture ibuprofen involved a lot of chemical reactions and produced a lot of waste. The atom economy was just 32%.
Recently it became possible for any pharmaceutical (drug) company to make ibuprofen. As there was a lot of money to be made, the race was on to find the most economic way to make it. This meant cutting down waste. The new method involves catalysts, some of which can be completely recovered and do not go out as waste. The atom economy is increased to 77%, partly because only the active form of ibuprofen is made. This also means that lower doses are needed and they take a shorter time to kill any pain.

Evaluate the two methods of manufacture in terms of the social, economic and environmental issues involved.
[Higher] (6)

HOW SCIENCE WORKS QUESTIONS

A class of students were given the task of finding out how much hydrogen would be produced by different amounts of calcium reacting with water. The hydrogen was collected in an upturned measuring cylinder. The apparatus was set up as is shown.

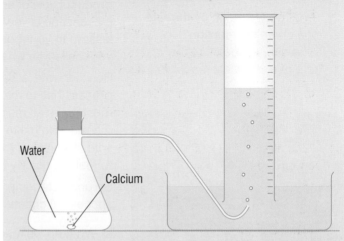

Water

Calcium

Different amounts of calcium were weighed . Each piece was put separately into the flask and the bung put on as quickly as possible. The reaction was left to finish and the volume of hydrogen measured.

The results of the different groups are in the table below.

Mass of Ca (g)	Volume of hydrogen (cm³)				
	A	B	C	D	E
0.05	25	26	27	18	26
0.10	55	53	55	55	52
0.15	85	86	81	89	84
0.20	115	117	109	116	113
0.25	145	146	148	141	140

a) Produce a table to show the mean for all of the groups for each mass of calcium used. (5)

b) Why is the mean often a more useful set of results than any one group? (1)

c) What is the range of volumes when 0.25 grams of calcium are used? (1)

d) What is the sensitivity of the balance used to weigh out the calcium? (1)

e) How could you determine which group had the most accurate results? (1)

f) Look at the method described Is there any possibility of a systematic error? If so, say how this error could arise. (1)

C2 4.1   How fast?

Figure 1 All living things depend on very precise control of the millions of chemical reactions happening inside their cells

The steeper the line on a graph, the faster the reaction rate.

The rate of a chemical reaction tells us how fast the reactants are turned into products. It is often very important for us to know about this. In a science class, if a reaction is very slow you won't get your results before the end of the lesson! In your body, chemical reactions must take place at rates which supply your cells with what they need exactly when they need them.

Reaction rate is also very important in the chemical industry. Any industrial process has to make money by producing useful products. This means we must make as much of the product we want as cheaply as possible. If it costs too much to make a chemical, it will be hard to make much profit when it is sold. The rate of the reaction used to make our chemical must be fast enough to make it quickly and safely.

So understanding and controlling reaction rates is always necessary for successful chemistry – whether in a cell or a chemical factory!

a) What do we mean by the *rate* of a chemical reaction?
b) Why is understanding the rate of chemical reactions so important?

How can we measure the rate of chemical reactions?

Chemical reactions happen at all sorts of different rates. Some are astonishingly fast – think of a firework exploding! Others are very slow, like rusting – the reaction between iron and oxygen in damp conditions.

There are two ways we can measure the rate of a chemical reaction. We can measure how quickly the reactants are used up as they react to make products. Or we can measure the rate at which the products of the reaction are made.

There are three main ways we can make these kinds of measurements.

PRACTICAL

Measuring the mass of a reaction mixture

We can measure the rate at which the **mass** of the reaction mixture changes if a gas is given off.

As the reaction takes place the mass of the reaction mixture will decrease. We can measure and record this at time intervals very easily.

Some balances can be attached to a computer to monitor the loss in mass continuously.

- Why is the cotton wool placed in the neck of the conical flask?
- How would the line on the graph differ if you plot 'Loss in mass' on the vertical axis?

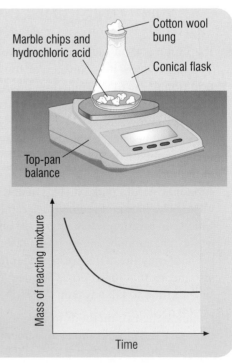

PRACTICAL

PRACTICAL

Measuring the volume of gas given off

If a gas is produced in a reaction, we can measure the rate of reaction. We do this by collecting the gas and measuring its volume at time intervals.

● What are the sources of error when measuring the volume of gas?

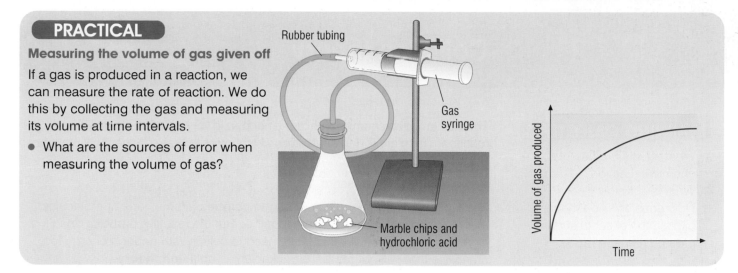

Rubber tubing

Gas syringe

Marble chips and hydrochloric acid

Volume of gas produced

Time

PRACTICAL

Measuring the light transmitted through a solution

Some reactions make an insoluble solid (precipitate) which makes the solution go cloudy. We can measure the rate at which the solid appears.

If the reaction is set up in a flask under which we put on a piece of paper marked with a cross, we can record the time taken for the cross to disappear. The shorter the time, the faster the reaction rate.

Or we can use a light meter connected to a data logger to measure the amount of light that can get through the solution, as the graph shows.

● What are the advantages of using a light meter rather than using the 'disappearing cross' method?

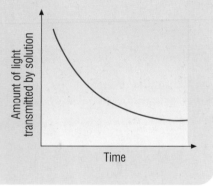

Amount of light transmitted by solution

Time

We can summarise these methods of working out the rate of a reaction using this equation:

$$\text{Rate of reaction} = \frac{\text{amount of reactant used or amount of product formed}}{\text{time}}$$

SUMMARY QUESTIONS

1 Copy and complete using the words below:

   explosion   products   rate   reactants   rusting

   Measuring the rate at which …… are used up or …… are made are two ways of measuring the …… of a chemical reaction. An example of a reaction that happens quickly is an …… . A reaction that happens slowly is …… .

2 Sketch graphs to show the results of the two main ways in which we can measure the rate of a chemical reaction. (See the first sentence in question 1 above.)

3 If we measure the time taken for a solution to become cloudy (see above) how is the time taken for the cross on the paper to disappear related to the rate of the chemical reaction?

KEY POINTS

1 Knowing and controlling the rate of chemical reactions is important in living cells, in the laboratory and in industry.

2 We can measure the rate of a chemical reaction by following the rate at which reactants are used up. Alternatively, we can measure the rate at which products are made.

C2 4.2 — Collision theory

The rate at which chemical reactions happen is very different, depending on the reaction. There are four main factors which affect the rate of chemical reactions:

- temperature,
- concentration or pressure,
- surface area,
- presence of a catalyst.

Reactions can only take place when the particles (atoms, ions or molecules) that make up the reactants come together. But the reacting particles don't just have to bump into each other. They need to collide with enough energy too, otherwise they will not react. This is known as **collision theory**.

The smallest amount of energy that particles must have before they will react is known as the **activation energy**. So anything which increases the chance of reacting particles bumping into each other, or which increases the energy that they have when they collide will make it more likely that reactions will happen. If we increase the chance of individual particles reacting, we will also increase the rate of reaction.

a) What must happen before two particles can stand a chance of reacting?
b) Particles must have a certain amount of energy before they will react – what is this energy called?

In everyday life we control the rates of chemical reactions often without any idea what we are doing and why! For example, cooking cakes in ovens or spraying a mixture of fuel and air into our car engines. But in chemistry we need to know exactly how to control the rate of chemical reactions and why our method works.

Surface area and reaction rate

If we want to light a fire we don't pile large logs together and try to set them alight. We use small pieces of kindling to begin with. Doing this increases the surface area of the logs, so there is more wood that can react with the air.

When a solid reactant reacts with a solution, the size of the pieces of the solid material make a big difference to the rate of the reaction. The inside of a large piece of solid is not in contact with the solution it is reacting with, so it can't react. It has to wait for the outside to react first.

In smaller lumps, or in a powder, each tiny piece is surrounded by solution. This means that reactions can take place much more easily.

c) How does the surface area of a solid affect its rate of reaction?

Figure 1 There is no doubt that the chemicals in these fireworks have reacted – but how do we explain what happens in a chemical reaction?

Figure 2 Cooking – an excellent example of controlling reaction rates!

Figure 3 When a solid reacts, the size of its pieces make a big difference to the rate of the reaction – the smaller the pieces, the faster the reaction

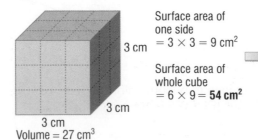

Surface area of one side = $3 \times 3 = 9 \text{ cm}^2$

Surface area of whole cube = $6 \times 9 = \textbf{54 cm}^2$

3 cm
3 cm
3 cm
Volume = 27 cm³

27 →

27 of these small cubes have the same volume as the large cube

1 cm
1 cm
1 cm

Surface area of one side = $1 \times 1 = 1 \text{ cm}^2$

Surface area of whole cube = $6 \times 1 = 6 \text{ cm}^2$

Surface area of 27 small cubes = $27 \times 6 = \textbf{162 cm}^2$

PRACTICAL

Which burns faster – ribbon or powder?

Make sure you have a heatproof mat under the Bunsen burner and you must wear goggles. Try lighting a 2 cm length of magnesium ribbon and time how long it takes to burn. Take a small spatula tip of magnesium powder and sprinkle it into the Bunsen flame.

● What safety precautions should you take in this experiment?

PRACTICAL

Investigating surface area

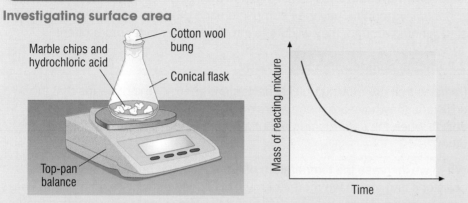

Marble chips and hydrochloric acid

Cotton wool bung

Conical flask

Top-pan balance

Mass of reacting mixture

Time

You can investigate the effect of changing surface area by measuring the mass lost against time for different sizes of marble chips. You need at least two different sizes of marble chips. These should be washed in dilute acid, rinsed with water and dried before they are used (to remove any powder on the surface).

● What variables should you control to make this a fair test?

A data logger would help to plot graphs of the results.

GET IT RIGHT!

Particles collide all the time, but only some collisions lead to reactions.

Increasing the number of collisions and the energy of collisions produces faster rates.

Larger surface area does not result in collisions with more energy but does increase the frequency of collisions.

SUMMARY QUESTIONS

1 Match the sentence fragments:

a) For two particles to react ……	A …… only if they have enough energy.
b) Two particles will react ……	B …… is called the activation energy.
c) The energy required for particles to react ……	C …… increases the rate at which it will react.
d) Increasing the surface area of a solid ……	D …… they must first collide.

2 Draw a diagram to explain why it is easier to light a fire using small pieces of kindling rather than big logs.

3 Why do you digest your food more quickly if you chew it well before you swallow it?

KEY POINTS

1 The minimum amount of energy that particles must have in order to react is called the activation energy.

2 The rate of a chemical reaction increases if the surface area of any solid reactants is increased.

C2 4.3 The effect of temperature

LEARNING OBJECTIVES

1 How does changing temperature affect the rate of reactions?

Figure 1 Moving faster means it's more likely that you'll bump into someone else – and the bump will be harder too!

When we increase the temperature of a reaction it always increases the rate of the reaction. Collision theory tells us why this happens – there are two reasons.

Both of these reasons are related to the fact that when we heat up a mixture of reactants the particles in the mixture move more quickly.

1 Particles collide more often

When we heat up a substance, energy is transferred to the particles that make up the substance. This means that they move faster. And when particles move faster they have more collisions. Imagine a lot of people walking around in the school playground. They may bump into each other occasionally – but if they start running around, they will bump into each other even more often!

2 Particles collide with more energy

Particles that are moving quickly have more energy, which means that the collisions they have are much more energetic. It's just like two people who collide when they are running about as opposed to just walking into each other!

When we increase the temperature of a reaction, the particles have more collisions and they have more energy. This speeds up the reaction in two ways – the particles will collide more often and they have more energy when they do collide.

Both of these changes increase the chance that two molecules will react. Around room temperature, if we increase the temperature of the reaction by 10°C the rate of the reaction will roughly double.

a) Why does increasing the temperature increase the rate of a reaction?
b) How much does a 10°C rise in temperature increase reaction rate at room temperature?

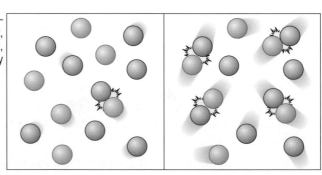

Cold – slow movement, few collisions, little energy

Hot – fast movement, more collisions, more energy

Figure 2 More collisions with more energy – both of these increase the rate of a chemical reaction as the temperature increases

This change in reaction rate is why we use fridges and freezers – because reducing the temperature slows down the rate of a chemical reaction. When food goes bad it is because of chemical reactions. Reducing the temperature slows down these reactions, so the food goes off much less quickly.

PRACTICAL

The effect of temperature on rate of reaction

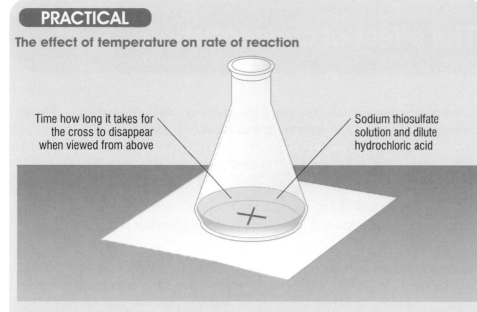

Time how long it takes for the cross to disappear when viewed from above

Sodium thiosulfate solution and dilute hydrochloric acid

NEXT TIME YOU...

. . . turn the heat up when you're cooking a meal, remember that you're increasing the rate at which chemical reactions are happening!

When we react sodium thiosulfate solution and hydrochloric acid it produces sulfur. This makes the solution go cloudy. We can record the length of time it takes for the solution to go cloudy at different temperatures.

● Which variables do you have to control to make this a fair test?
● Why is it difficult to get accurate timings by eye in this investigation?
● How can you improve the reliability of the data you collect?

The results of an investigation like this can be plotted on a graph:

The graph shows how the time for the solution to go cloudy changes with temperature.

c) What happens to the time it takes the solution to go cloudy as the temperature increases?

Time taken for solution to go cloudy

Temperature

SUMMARY QUESTIONS

1 Copy and complete using the words below:

**chemical   collide   decreases   doubles   energy   off
quickly   rate   reducing   rise**

When we increase the temperature of a reaction, we increase its .... This makes the particles move more ...... so they ...... more often and they have more .... At room temperature, a temperature ...... of about 10°C roughly ...... the reaction rate. This explains why we use fridges and freezers – because ...... the temperature ...... the rate of the ...... reactions which make food go ....

2 Water in a pressure cooker boils at a much higher temperature than water in a saucepan because it is under pressure. Why does food take longer to cook in a saucepan than it does in a pressure cooker?

3 Use your knowledge of the effect of temperature on chemical reactions to explain why cold-blooded animals like reptiles or insects may move very slowly in cold weather.

KEY POINTS

1 Reactions happen more quickly as the temperature increases.
2 A 10°C increase in temperature at room temperature roughly doubles the rate of a reaction.
3 The rate of a chemical reaction increases with temperature because the particles collide more often and they have more energy.

C2 4.4   The effect of concentration

LEARNING OBJECTIVES

1 How does changing the concentration of reactants affect the rate of reactions?
2 How does changing the pressure of reacting gases affect the rate of reactions?

Some of our most beautiful buildings are made of limestone or marble. These buildings have stood for centuries, but in the last 50 years or so they have begun crumbling away increasingly fast. This is because limestone and marble both contain calcium carbonate. This reacts with acids, leaving the stone soft and crumbly.

We think that the rate of this reaction has speeded up because the concentration of sulfuric and nitric acids found in rainwater has been steadily increasing.

Increasing the concentration of reactants in a solution, increases reaction rate because there are more particles of the reactants moving around in the same volume. The more 'crowded' together the reactant particles are, the more likely it is that they will bump into each other and a reaction will take place.

Increasing the pressure of a reaction involving gases has the same effect. It squashes the gas particles more closely together. This increases the chance that they will collide and react and so speeds up the rate of the reaction.

a) Why does increasing concentration or pressure increase reaction rate?

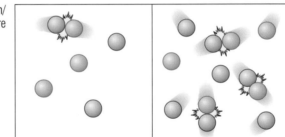

Low concentration/ low pressure   High concentration/ high pressure

Figure 2 Increasing concentration and pressure both mean that particles are closer together. This increases the number of collisions between particles, so the reaction rate increases.

Figure 1 Limestone statues are damaged by acid rain. This damage increases as the concentration of the acids in rainwater increases.

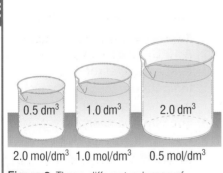

0.5 dm³   1.0 dm³   2.0 dm³

2.0 mol/dm³ 1.0 mol/dm³   0.5 mol/dm³

Figure 3 These different volumes of solution all contain the same amount of solute – but at different concentrations

The concentration of a solution tells us how many particles of solute we have dissolved in a certain volume of the solution. Concentration is measured in moles per cubic decimetre, which is shortened to mol/dm^3. Solutions with the same concentration always contain the same number of particles of solute in the same volume.

b) What unit do we use to measure the concentration of solute in a solution?

We never talk about the concentration of a gas – but the number of particles in a certain volume of gas depends on its temperature and its pressure. At the same temperature and pressure, equal volumes of gases all contain the same number of particles.

c) Two identical containers of gas are at the same temperature and pressure. What can we say about the number of particles in the two containers?

PRACTICAL

Investigating the effect of concentration on rate of reaction

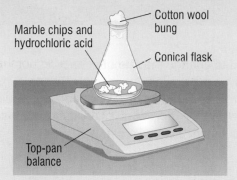

Marble chips and hydrochloric acid

Cotton wool bung

Conical flask

Top-pan balance

We can investigate the effect of changing concentration by reacting marble chips with different concentrations of hydrochloric acid, which produces carbon dioxide gas:

$$CaCO_3 + 2HCl \rightarrow CaCl_2 + CO_2 + H_2O$$

We can measure the rate of reaction by plotting the mass of the reaction mixture as carbon dioxide gas is given off in the reaction.

- How do you make this a fair test?
- What conclusion can you draw from your results?

If we plot the results of an investigation like the one above on a graph they look like this:

The graph shows how the rate at which the mass of the reaction mixture decreases changes with concentration.

d) Which line on the graph shows the fastest reaction? How could you tell?

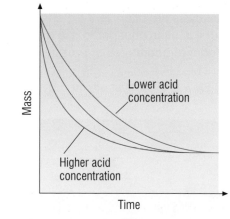

Mass

Lower acid concentration

Higher acid concentration

Time

SUMMARY QUESTIONS

1 Copy and complete using the words below:

> **collisions   concentration   faster   gases   increases**
> **number   pressure   rate   volume**

The ...... of a chemical reaction is affected by the ...... of reactants in solution and by ...... if the reactants are .... Both of these tell us the ...... of particles that there are in a certain ...... of the reaction mixture. Increasing this ...... the number of ...... that particles make with each other, making reactions happen ....

2 Acidic cleaners are designed to remove limescale when they are used neat. They do not work so well when they are diluted. Using your knowledge of collision theory, explain why this is.

3 How are the 'concentration of a solution' and the 'pressure of a gas' similar? [Higher]

KEY POINTS

1 Increasing the concentration of reactants increases the frequency of collisions between particles, increasing the rate of reaction.

2 Increasing the pressure of reacting gases results in particles colliding more often, increasing the rate of reaction.

C2 4.5   The effect of catalysts

Sometimes we need to change the rate of a reaction but this is impossible using any of the ways we have looked at so far. Or sometimes a reaction might be possible only if we use very high temperatures or pressures – which can be very expensive. However we can speed chemical reactions up another way – by using a special substance called a **catalyst**.

a) Apart from using a catalyst, what other ways are there of speeding up a chemical reaction?

A catalyst is a substance which increases the rate of a chemical reaction but it is not affected chemically itself at the end of the reaction. It is not used up in the reaction, so it can be used over and over again to speed up the conversion of reactants to products.

We need to use different catalysts with different reactions. Many of the catalysts we use in industry are transition metals or their compounds. For example, iron is used in the Haber process, while platinum is used in the production of nitric acid.

Catalysts are often very expensive because they are made of precious metals. But it is often cheaper to use a catalyst than to pay for all the energy needed for higher temperatures or pressures in a reaction.

b) How is a catalyst affected by a chemical reaction?

Figure 1 Catalysts are all around us, in the natural world and in industry. Our planet would be very different without them.

Figure 2 The transition metals platinum and palladium are used in the catalytic converters in cars

Some catalysts work by providing a surface for the reacting particles to come together. They lower the activation energy needed for the particles to react. This means that more of the collisions between particles result in a reaction taking place. We normally use catalysts in the form of powders, pellets or fine gauzes. This gives the biggest possible surface area for them to work.

c) Why is a catalyst divided up into pellets more effective than a whole lump of the catalyst?

PRACTICAL
Investigating catalysis

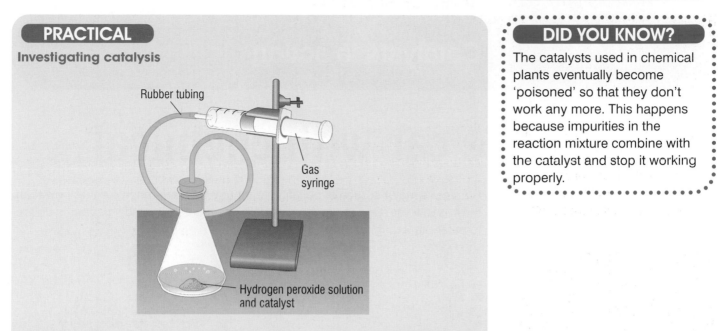

Rubber tubing

Gas syringe

Hydrogen peroxide solution and catalyst

We can investigate the effect of different catalysts on the rate that hydrogen peroxide solution decomposes:

$$2H_2O_2 \rightarrow 2H_2O + O_2$$

The reaction produces oxygen. We can collect this in a gas syringe using the apparatus shown above.

We can investigate the effect of many different substances on the rate of this reaction. Examples include manganese(IV) oxide and potassium iodide.

- State the independent variable in this investigation. (See page 2.)

A simple table of the time taken to produce a certain volume of oxygen can then tell us which catalyst makes the reaction go fastest.

- What type of graph would you use to show the results of your investigation? Why? (See page 13.)

Apart from speeding up a chemical reaction, the most important thing about a catalyst is that it does not get used up in the chemical reaction. We can use a tiny amount of catalyst to speed up a chemical reaction over and over again.

SUMMARY QUESTIONS

1 Copy and complete using the words below:

   activation   energy   increases   more   react

   A catalyst ...... the rate of a chemical reaction. It does this by reducing the ...... energy needed for the reaction. This means that ...... particles have enough ...... to ....

2 Solid catalysts used in chemical plants are often shaped as tiny beads or cylinders with holes through them. Why are they made in this shape?

3 Why is the number of moles of catalyst needed to speed up a chemical reaction very small compared to the number of moles of reactants?

KEY POINTS

1 A catalyst speeds up the rate of a chemical reaction.
2 A catalyst is not used up during a chemical reaction.

C2 4.6   Catalysts in action

Cleaning the car with chemical catalysts

Cars are a major source of pollution, although they are much cleaner now than they used to be. One reason that petrol-fuelled cars are much cleaner is down to catalysts . . .

Fuel travels to the engine. Here it is mixed with air and passes into the cylinders. At just the right point the petrol and air mixture is made to explode by a tiny electric spark. This explosion provides the force that pushes the piston downwards to make the car move. The explosion makes carbon dioxide and water as the hydrocarbon reacts with oxygen in the air. But carbon monoxide and nitrogen oxides are made too. Carbon monoxide is toxic and nitrogen oxides contribute towards acid rain.

Petrol, a fuel made of hydrocarbons, goes into the tank. Lead compounds used to be added to the petrol to improve the performance of the engine. Lead is no longer added to petrol as it is poisonous.

The exhaust gases from the engine pass out through the exhaust pipe and through a catalytic converter. Here the gases pass over a metal catalyst. This removes the oxygen from the nitrogen oxides and reacts it with the carbon monoxide. The result is carbon dioxide and nitrogen.

The catalyst used may be platinum, palladium or rhodium, or a combination of these transition metals. The catalyst is arranged so that it has a very large surface area. Catalysts can be 'poisoned' by lead in petrol – so it is very important to use 'unleaded' petrol in a car that is fitted with a catalytic converter.

ACTIVITY

The diagram shows some of the chemistry that goes to make cars much 'cleaner' than they used to be.

Write a short article for the motoring section of a local newspaper describing why a modern car causes much less pollution than a car built thirty years ago. Remember that the readers may not have much scientific knowledge, so any chemistry will need to be explained using simple language.

Remember to think of a catchy title for your article!

ENZYMES – CLEVER CATALYSTS THAT ARE GETTING EVERYWHERE

ENZYMES MAKE CLOTHES CLEANER AND CLASSIER!

For years we've been used to enzymes helping to get our clothes clean. Biological washing powders contain tiny molecules that help to literally 'break apart' dirt molecules such as proteins at low temperatures. But why stop there? Why not make washing powders that help to repair clothes and make them like new? Searching the surface of fabric for any tears or breaks in the fibres, enzymes could join these back together, while other enzymes might look for frayed or 'furry' bits of fabric and could makes these smooth again. And not only that . . . . . .

ENZYMES TO THE RESCUE

Everyone knows how upsetting it is to cut yourself. And it can be a pain to have to wear a sticking plaster until your body has repaired your skin. But enzymes may be the answer to this. By choosing the right enzyme mixture we may one day be able to mend cuts and other damage to our skin simply by painting a liquid onto our skin. But until then . . . . . .

SAY 'AAAHH' FOR THE ENZYME!

Those lengthy waits for the results of a blood test to come back could soon be a thing of the past. By combining biological molecules like enzymes with electronics, scientists reckon that they can make tiny measuring probes that will enable doctors to get an instant readout of the level of chemicals in your blood. These sensors are so tiny that it is possible to take more than a dozen measurements at the same time, doing away with the need for lots of tests.

SCIENTISTS MAKE ENZYMES MAKE COMPUTER

Scientists announced today that they are close to making a biological computer made of enzymes and DNA. The tiny device could change the face of computing in the future, which up until now has been based on electronic devices made from silicon. The idea of using DNA in computers first took off in 1994, when a scientist in California used it to solve a maths problem. Computers made from DNA would be so tiny that a trillion of them could fit in a single drop of water.

ACTIVITY

There is more than a grain of truth in all of these ideas about enzymes. Use some or all of these news reports to produce a short piece for the 'and finally' slot in a TV news broadcast. Remember to use your knowledge of catalysts to explain why enzymes are important in all of these developments.

SUMMARY QUESTIONS

1 Select from A, B and C to show how the rate of each reaction, a) to d), could be measured.

a) Gas evolved from reaction mixture

b) Mass of reaction mixture changes

c) Precipitate produced

d) Colour of solution changes

A Measure mass

B Measure light transmitted

C Measure volume

2 A student carried out a reaction in which she dropped a piece of magnesium ribbon in sulfuric acid with a concentration of 1 mol/dm³.

a) Suggest **one** way in which the student could measure the rate of this reaction.

b) Suggest **three** ways in which the student could increase the rate of this reaction.

c) Explain how each of these methods changes the rate of the reaction.

3 The following results show what happened when two students investigated the reaction of some marble chips with acid.

Time (seconds)	Investigation 1 Volume of gas produced (cm³)	Investigation 2 Volume of gas produced (cm³)
0	0	0
30	5	10
60	10	20
90	15	30
120	20	40
150	25	50
180	28	57
210	30	59
240	30	60

a) Plot a graph of these results with time on the *x*-axis.

b) After 30 seconds, how does the rate of the reaction in investigation 2 compare with the rate of reaction in investigation 1?

c) How does the final volume of gas produced in investigation 2 compare with the final volume of gas produced in investigation 1?

d) Suggest how the reaction in investigation 2 differs from the reaction in investigation 1. Explain your answer.

EXAM-STYLE QUESTIONS

1 Marble chips (calcium carbonate) react with hydrochloric acid as shown in the equation.

$$CaCO_3(s) + 2HCl(aq) \rightarrow CaCl_2(aq) + H_2O(l) + CO_2(g)$$

Some students investigated the effect of the size of marble chips on the rate of this reaction. They did the reactions in a conical flask, which they put onto a balance connected to a computer to record their results. They used three different sizes of marble chips and kept all of the other conditions the same. The graphs show the total mass of the flask and reaction mixture plotted against time for the three experiments.

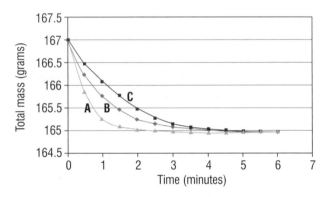

(a) Which curve, **A**, **B** or **C**, shows the results for the fastest reaction? (1)

(b) Which curve, **A**, **B** or **C**, shows the results for the largest marble chips? (1)

(c) Explain, using collision theory, why changing the size of marble chips changes the rate of reaction. (2)

(d) (i) Use curve **A** to describe how the rate of reaction changes from the start to the finish of the reaction. (3)

(ii) Explain why the rate of reaction changes in this way. (2)

2 A student investigated the reaction between magnesium ribbon and hydrochloric acid.

$$Mg(s) + 2HCl(aq) \rightarrow MgCl_2(aq) + H_2(g)$$

The student reacted 20 cm³ of two different concentrations of hydrochloric acid with 0.050 g of magnesium. All other conditions were kept the same. The student's results are shown in the table on the next page.

Concentration of acid (moles per dm³)	Time (minutes)	0	1	2	3	4	5	6	7	8	9	10
1.0	Volume of gas (cm³)	0	15	24	31	37	41	44	46	47	48	48
2.0	Volume of gas (cm³)	0	30	39	45	47	48	48	48	48	48	48

(a) Name the dependent variable. (1)

(b) Suggest a control variable. (1)

(c) Suggest how the student might have controlled this variable. (1)

(d) Plot these results on the same axes, with time on the horizontal axis and volume of gas on the vertical axis. Draw a smooth line for each concentration. Label each line with the concentration of acid. (4)

(e) (i) What is the effect of doubling the concentration on the rate of reaction? (1)

(ii) Explain how the graphs show this effect. (2)

(iii) Explain this effect in terms of particles and collision theory. (2)

(f) Explain why the total volume of hydrogen is the same for both reactions. (2)

(g) Draw a labelled diagram of the apparatus you would use to do this experiment. (3)

3 Hydrogen peroxide solution is colourless and decomposes very slowly at 20°C.

$$2H_2O_2(aq) \rightarrow 2H_2O(l) + O_2(g)$$

Manganese(IV) oxide, a black powder, is a catalyst for the reaction.

(a) Explain what the word catalyst means. (2)

(b) What would you **see** if manganese(IV) oxide was added to hydrogen peroxide solution? (1)

(c) Describe briefly one way that you could show that manganese(IV) oxide was acting as a catalyst. (2)

(d) Explain, using particle and collision theory, how a solid catalyst works. (2)

(e) Hydrogen peroxide solution stored at 10°C decomposes at half the rate compared to when it is stored at 20°C. Explain, in terms of particles, why the rate of the reaction changes in this way. (3)

HOW SCIENCE WORKS QUESTIONS

This student's account of an investigation into the effect of temperature on the rate of a reaction was found on the Internet.

I investigated the effect of temperature on the rate of a reaction. The reaction was between sodium thiosulfate and hydrochloric acid. I set up my apparatus as in this diagram.

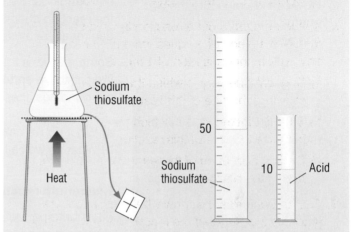

The cross was put under the flask. I heated the sodium thiosulfate to the temperature I wanted and then added the hydrochloric acid to the flask. I immediately started the watch and timed how long it took for the cross to disappear.

My results are below.

Temperature of the sodium thiosulfate	Time taken for the cross to disappear
15	110
30	40
45	21

My conclusion is that the reaction goes faster the higher the temperature.

a) Suggest a suitable prediction for this investigation. (1)

b) Describe one safety feature that is not mentioned in the method. (1)

c) Suggest some ways in which this method could be improved. For each suggestion, say why it is an improvement. (10)

d) Suggest how the table of results could be improved. (1)

e) Despite all of the problems with this investigation, is the conclusion appropriate? Explain your answer. (1)

C2 5.1 Exothermic and endothermic reactions

Figure 1 When a fuel burns in oxygen, energy is transferred to the surroundings. We usually don't need a thermometer to know that there is a temperature change!

Whenever chemical reactions take place, energy is involved. That's because energy is always transferred as chemical bonds are broken and formed.

Some reactions transfer energy *from* the reacting chemicals *to* the surroundings. We call these **exothermic** reactions. The energy transferred from the reacting chemicals often heats up the surroundings. This means that we can measure a rise in temperature as the reaction happens.

Some reactions transfer energy *from* the surroundings *to* the reacting chemicals. We call these **endothermic** reactions. Because they take in energy from their surroundings, these reactions cause a drop in temperature as they happen.

a) What do we call a chemical reaction that gives out heat?
b) What do we call a chemical reaction that absorbs heat from its surroundings?

Exothermic reactions

Fuels burning are an obvious example of exothermic reactions, but there are others which we often meet in the chemistry lab.

Neutralisation reactions between acids and alkalis are exothermic. We can easily measure the rise in temperature using simple apparatus (see opposite).

Similarly, heat is released when we add water to white anhydrous copper(II) sulfate (anhydrous means 'without water'). This reaction makes blue hydrated copper(II) sulfate crystals. The reaction gives out heat – it is an exothermic reaction.

Respiration is a very special kind of burning. It involves reacting sugar with oxygen inside the cells of every living thing. This makes the energy needed for all the reactions of life, and also makes water and carbon dioxide as waste products. Respiration is another exothermic reaction.

c) Give two examples of exothermic reactions.

Figure 2 All warm-blooded animals rely on exothermic reactions to keep their body temperatures steady

Endothermic reactions

Endothermic reactions are much less common than exothermic ones.

When we dissolve some ionic compounds like potassium chloride or ammonium nitrate in water the temperature of the solution drops.

Thermal decomposition reactions are also endothermic. An example is the decomposition of calcium carbonate to form calcium oxide and carbon dioxide. This reaction only takes place if we keep heating the calcium carbonate strongly. It takes in a great deal of energy from the surroundings.

The most important endothermic reaction of all is **photosynthesis**. This is the reaction in which plants turn carbon dioxide and water into sugar and oxygen, using energy from the Sun.

Figure 3 When we eat sherbet we can feel an endothermic reaction! Sherbet dissolving in the water in your mouth takes in energy – giving a slight cooling effect.

d) Name two endothermic reactions.

PRACTICAL

Investigating energy changes

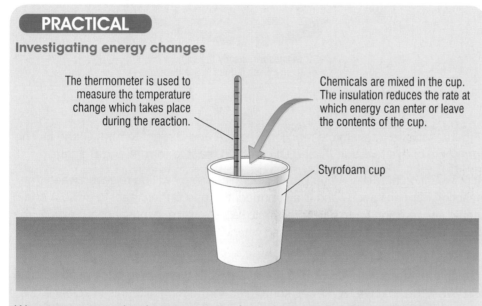

The thermometer is used to measure the temperature change which takes place during the reaction.

Chemicals are mixed in the cup. The insulation reduces the rate at which energy can enter or leave the contents of the cup.

Styrofoam cup

We can use very simple apparatus to investigate the energy changes in reactions. Often we don't need to use anything more complicated than a Styrofoam drinks cup and a thermometer.

- State two ways in which you could make the data you collect more reliable.

SUMMARY QUESTIONS

1 Copy and complete using the words below:

**broken   endothermic   exothermic   made   neutralisation
photosynthesis   respiration   thermal decomposition**

Chemical reactions involve energy changes as bonds are ...... and .... When a chemical reaction releases energy we say that it is an ...... reaction. Two important examples of this kind of reaction are ...... and .... When a chemical reaction takes in energy we say that it is an ...... reaction. Two important examples of this kind of reaction are ...... and ....

2 Potassium chloride dissolving in water is an endothermic process. What might you expect to observe when potassium chloride dissolves in water?

KEY POINTS

1 Energy may be transferred to or from the reacting substances in a chemical reaction.
2 A reaction where energy is transferred from the reacting substances is called an exothermic reaction.
3 A reaction where energy is transferred to the reacting substances is called an endothermic reaction.

C2 5.2   Energy and reversible reactions

LEARNING OBJECTIVES

1 How is energy involved in reversible reactions?
2 What happens to reactions at equilibrium when we change the temperature? [Higher]

Energy changes are involved in reversible reactions too. We can understand energy changes in a reversible reaction if we think carefully about the reaction itself.

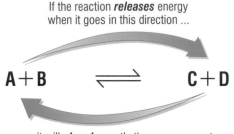

If the reaction **releases** energy when it goes in this direction ...

$$A+B \rightleftharpoons C+D$$

... it will **absorb** exactly the same amount of energy when it goes in this direction

Figure 1 A reversible reaction

Figure 1 shows a reversible reaction where A and B react to form C and D. The products of this reaction (C and D) can then react to form A and B again.

If the reaction between A and B is exothermic, energy will be released when the reaction happens and C and D are formed. If C and D then react to make A and B again, the reaction must be endothermic. What's more, it must absorb exactly the same amount of energy as it released when C and D were formed from A and B.

If this didn't happen it would be possible for us to 'make' energy just by continuously reacting A and B to make C and D, then going back again and so on. We know that we never get 'something for nothing' like this in science. So the amount of energy released when we go in one direction in a reversible reaction must be exactly the same as the energy absorbed when we go in the opposite direction.

a) How does the energy change for a reversible reaction in one direction compare to the energy change for the reaction in the opposite direction?

We can see how this works if we look at what happens when we heat blue copper sulfate crystals. The crystals contain water as part of the lattice formed when they crystallise. We say that the copper sulfate is **hydrated**. Heating the copper sulfate drives off the water from the crystals, producing white **anhydrous** ('without water') copper sulfate. This is an endothermic reaction.

$$\underset{\substack{\text{hydrated}\\\text{copper sulfate}}}{\underset{\text{blue crystals}}{CuSO_4.5H_2O}} \rightleftharpoons \underset{\substack{\text{anhydrous}\\\text{copper sulfate}}}{\underset{\text{white powder}}{CuSO_4}} + 5H_2O$$

When we add water to anhydrous copper sulfate we form hydrated copper sulfate. The colour change in the reaction is a useful test for water. The reaction in this direction is exothermic. In fact, so much energy may be produced that we may see steam rising as water boils.

PRACTICAL

Energy changes in a reversible reaction

Try these reactions yourself. Gently heat a few copper sulfate crystals in a test tube. Observe the changes. When the crystals are completely white allow the tube to cool to room temperature (takes several minutes). Add two or three drops of water from a dropper and observe the changes. Carefully feel the bottom of the test tube.

- Explain the changes you have observed.

You can repeat this with the same solid, as it is a reversible reaction or try with other hydrated crystals, like cobalt chloride. Some are not so colourful but the changes are similar.

b) What can anhydrous copper sulfate be used to test for?

HIGHER

Energy and equilibrium

We have a closed system when nothing is added or taken away from the reaction mixture. In a closed system the relative amounts of the reactants and products in a reversible reaction depend on the temperature.

The ability to change the balance of reactants and products is very important when we look at industrial processes. By changing the temperature we can get more of the products and less of the reactants. Look at the table:

If a reaction is exothermic	If a reaction is endothermic ......
... an increase in temperature decreases the yield of the reaction, so the amount of products formed is lower.	... an increase in temperature increases the yield of the reaction, so the amount of products formed is larger.
... a decrease in temperature increases the yield of the reaction, so the amount of products formed is larger.	... a decrease in temperature decreases the yield of the reaction, so the amount of products formed is lower.

SUMMARY QUESTIONS

1 Copy and complete using the words below:

decreasing   endothermic   exothermic   increasing   reversible

If a reversible reaction is ...... in one direction it will be ...... in the opposite direction. If a ...... reaction is exothermic in the forward direction, ... the temperature will increase the amount of products formed. If it is endothermic in the forward direction, ... the temperature will increase the amount of products formed.   [Higher]

2 Blue cobalt chloride crystals turn pink when they become damp. The formula for the two forms can be written as $CoCl_2.2H_2O$ and $CoCl_2.6H_2O$.

a) How many moles of water will combine with 1 mole of $CoCl_2.2H_2O$?
b) Write a balanced equation for the reaction, which is reversible.
c) You have some *pink* cobalt chloride crystals. Suggest how you could turn these into *blue* cobalt chloride crystals.

KEY POINTS

1 In reversible reactions, one reaction is exothermic and the other is endothermic.
2 In any reversible reaction, the amount of energy released when the reaction goes in one direction is exactly equal to the energy absorbed when the reaction goes in the opposite direction.
3 We can change the amount of products formed at equilibrium by changing the temperature at which we carry out a reversible reaction. [Higher]

C2 5.3   More about the Haber process

LEARNING OBJECTIVES

1 Why do we use a temperature of 450°C for the Haber process?
2 Why do we use a pressure of about 200 atmospheres for the Haber process?

We saw on the previous page that the temperature at which we carry out a reversible reaction can affect the amount of the products formed at equilibrium. But if the reaction we are carrying out involves gases, pressure can be very important too.

Many reversible reactions which involve gases have more moles of gas on one side of the equation than on the other. By changing the pressure at which we carry out the reaction we can change the amount of products that we produce. Look at the table below:

If a reaction produces a larger volume of gases ……	If a reaction produces a smaller volume of gases ……
… an increase in pressure decreases the yield of the reaction, so the amount of products formed is lower.	… an increase in pressure increases the yield of the reaction, so the amount of products formed is larger.
… a decrease in pressure increases the yield of the reaction, so the amount of products formed is larger.	… a decrease in pressure decreases the yield of the reaction, so the amount of products formed is lower.

To see how this is useful we can look at the Haber process which we met earlier. (See pages 154 and 155.)

a) Look at the table above. How does increasing the pressure affect the amount of products formed in a reaction which produces a larger volume of gas?

The economics of the Haber process

The Haber process involves the reversible reaction between nitrogen and hydrogen to make ammonia:

$$N_2 + 3H_2 \rightleftharpoons 2NH_3 \text{ ($\rightleftharpoons$ is the equilibrium symbol)}$$

Energy is released during this reaction, so it is exothermic. As the chemical equation shows, there are 4 moles of gas ($N_2 + 3H_2$) on the left-hand side of the equation. But on the right-hand side there are only 2 moles of gas ($2NH_3$). This means that the volume of the reactants is much greater than the volume of the products. So an increase in pressure will tend to produce more ammonia.

b) How does the volume of the products in the Haber process compare to the volume of the reactants?

To get the maximum possible yield of ammonia in the Haber process, we need to make the pressure as high as possible. But high pressures need expensive reaction vessels and pipes which are strong enough to withstand the pressure. Otherwise there is always the danger that an explosion may happen.

In the Haber process we have to make a compromise between using very high pressures (which would produce a lot of ammonia) and the expense of building a chemical plant which can withstand those high pressures. This compromise means that we usually carry out the Haber process at between 200 and 350 atmospheres pressure.

Figure 1 It is expensive to build chemical plants that operate at high pressures

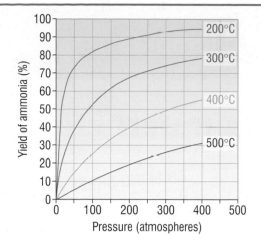

Figure 2 The conditions for the Haber process are a compromise between getting the maximum amount of product in the equilibrium mixture and getting the reaction to take place at a reasonable rate

The effect of temperature on the Haber process is more complicated than the effect of pressure. The forward reaction is exothermic. So if we carry it out at low temperature this would increase the amount of ammonia in the reaction mixture at equilibrium.

But at a low temperature, the rate of the reaction would be very slow. That's because the particles would collide less often and would have less energy. To make ammonia commercially we must get the reaction to go as fast as possible. We don't want to have to wait for the ammonia to be produced!

To do this we need another compromise. A reasonably high temperature is used to get the reaction going at a reasonable rate, even though this reduces the amount of ammonia in the equilibrium mixture.

We also use an iron catalyst to speed up the reaction. (Since this affects the rate of reaction in both directions, it does not affect the amount of ammonia in the equilibrium mixture.)

SUMMARY QUESTIONS

1 Copy and complete using the words below:

> **decreases   exothermic   fewer   increasing   left**
> **pressure   released**

The Haber process is …… so energy is …… during the reaction. This means that …… the temperature …… the amount of ammonia formed. Increasing the …… will increase the amount of ammonia formed, because there are …… moles of gas on the right-hand side of the equation than on the ……-hand side.

2 Look at Figure 2.

a) What is the approximate yield of ammonia at a temperature of 500°C and 400 atmospheres pressure?
b) What is the approximate yield of ammonia at a temperature of 500°C and 100 atmospheres pressure?
c) What is the approximate yield of ammonia at a temperature of 200°C and 400 atmospheres pressure?
d) What is the approximate yield of ammonia at a temperature of 200°C and 100 atmospheres pressure?
e) Why is the Haber process carried out at around 200 to 350 atmospheres and 450°C?

KEY POINTS

1 The Haber process uses a pressure of around 200 to 350 atmospheres to increase the amount of ammonia produced.
2 Although higher pressures would produce more ammonia, they would make the chemical plant too expensive to build.
3 A temperature of about 450°C is used for the reaction. Although lower temperatures would increase the amount of ammonia at equilibrium, the ammonia would be produced too slowly.

C2 5.4   Industrial dilemmas

How can we make as much chemical as possible . . .

ABC Laboratory Consultants

Haber House • Drudge Street • Anywhere • AD13 4FU

Dear Sirs

We are planning to build a factory to produce our new chemical, which has the secret formula AB. We are including some data sheets giving details of the reaction we shall be using to produce this chemical, and would like you to advise us about the best reaction conditions (temperature, pressure etc) to use to get as much AB as we can as cheaply as possible. We should like you to present your ideas in a short presentation to be held in your offices in two weeks' time.

Signed

BRIEFING SHEET 1

Project number: 45AB/L1670-J4550K
Specification: R MST3K 65 L7

Brief prepared by   J K Rolling
Checked by   L Skywalker

The equation for the reaction is:

$$A_2B_2 \rightleftharpoons 2AB$$

Both A_2B_2 and AB are gases. These are not their real formulae, which are secret. But the reaction does involve making two moles of product from one mole of reactants.

ACTIVITY

Working in teams, decide what you will advise Consolidated Chemicals to do about the conditions for the reaction. Prepare a presentation with your advice – the whole team should contribute to this. The following questions may help you:

● How does the amount of product change with temperature?
● How does the volume of the gases in the reaction change as AB is made from A_2B_2?
● What conditions may affect the reaction, and how?

BRIEFING SHEET 2

Project number: 45AB/L1670-J4550K
Specification: R MST3K 65 L7

Brief prepared by   J K Rolling
Checked by   L Skywalker

The graph shows the amount of AB in the equilibrium mixture at different temperatures.

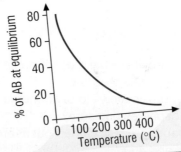

... and what happens when the raw materials run out?!

'What do we do when our resources run low?'

The world population grows all the time. It grows in its demands for a better lifestyle as well. Why shouldn't everyone have access to cars, computers and the latest electrical goodies? Yet all this growth means greater use of our natural resources – chemicals, minerals, oil. Minerals and metals don't replace themselves as carefully managed living resources do. So either we will have to find alternative materials, or alternative sources of the minerals we have been using . . .

London today saw some of the worst rioting as people struggled to get their hands on the last deliveries to be made to the shops. The world shortage of minerals has really begun to bite now, with the supplies of raw materials like copper and zinc running low and prices going through the roof. As oil supplies dwindle too, the lights are going out all over London . . .

'SPACE IS THE ANSWER!!'

The Bugle says 'get into space to find more minerals!!' It must be obvious even to our dim-witted leaders that we need to go and explore. Just as explorers in the past found new lands and new riches, we must go into space to find minerals on other planets! We can then bring them back to Earth so that we can make the things we need!'

ACTIVITY

There are many technical problems that have to be solved to allow us to travel to other planets in the Solar System. But imagine that they could be overcome. Could we really travel to other worlds to find new resources and bring them back to Earth?

Work in teams. You have been asked to produce a report for a government department about the possibility of using the Moon and nearby planets as a source of minerals. You need to consider not only the practical aspects of this but also the economics and the politics too. For example, in 1969 American astronauts landed on the Moon. So does this mean that the USA owns the Moon? Who will decide who owns the minerals on Mars or on the Moon?

SUMMARY QUESTIONS

1 Select from A, B, C and/or D to describe correctly exothermic and endothermic reactions.

a) In an exothermic reaction ......

b) In an endothermic reaction ......

A ...... we may notice a decrease in temperature.

B ...... energy is released by the chemicals.

C ...... we may notice an increase in temperature.

D ...... energy is absorbed by the chemicals.

2 'When sherbet sweets dissolve in your mouth this is an endothermic process.' Devise an experiment to test your statement. Use words and diagrams to describe clearly what you would do.

3 Two chemicals are mixed and react endothermically. When the reaction has finished, the reaction mixture is allowed to stand until it has returned to its starting temperature.

a) Sketch a graph of temperature (*y*-axis) *v* time (*x*-axis) to show how the temperature of the reaction mixture changes.

b) Label the graph clearly and explain what is happening wherever you have shown the temperature changing.

4 A chemical reaction can make product Z from reactants X and Y. Under the reaction conditions, X, Y and Z are gases.

X, Y and Z react in the proportions 1 : 2 : 3. The reaction is carried out at 250°C and 100 atmospheres. The reaction is reversible, and it is exothermic in the forward direction.

a) Write an equation for this (reversible) reaction.

b) How would increasing the pressure affect
   i) the amount of Z formed,
   ii) the rate at which Z is formed?

c) How would increasing the temperature affect
   i) the amount of Z formed,
   ii) the rate at which Z is formed?

d) A 10% yield of Z is obtained in 25 seconds under the reaction conditions. To get a 20% yield of Z under the same conditions takes 75 seconds. Explain why it makes more sense economically to set the reaction up to obtain a 10% yield rather than a 20% yield. [Higher]

EXAM-STYLE QUESTIONS

1 Match each of (a) to (g) with one of the following:

endothermic reaction   exothermic reaction
no reaction

(a) Burning petrol in a car engine.

(b) Respiration in living cells.

(c) Boiling water.

(d) Converting limestone into calcium oxide.

(e) Switching on an electric light bulb.

(f) Reducing lead oxide with carbon to produce lead.

(g) Carbon dioxide combining with water in cells of green plants. (7)

2 When heated continuously, pink cobalt chloride crystals can be changed into blue crystals .

$$CoCl_2.6H_2O \rightleftharpoons CoCl_2.2H_2O + 4H_2O$$
pink   blue

(a) What does the symbol $\rightleftharpoons$ tell you about this reaction? (1)

(b) How can you tell that the reaction to produce blue crystals is endothermic? (1)

(c) (i) How could you change the blue crystals to pink crystals? (1)

   (ii) What temperature change would you observe when this is done? (1)

(d) Suggest how the colour changes of these crystals could be used. (1)

3 The equation for the main reaction in the Haber process to make ammonia is:

$$N_2 + 3H_2 \rightleftharpoons 2NH_3$$

The table shows the percentage yield of the Haber process at different temperatures and pressures.

Pressure (atm)	Temp. (°C) 0	100	200	300	400	500
400	99	91	78	55	32	20
200	96	87	66	40	21	12
100	94	79	50	25	13	6
50	92	71	36	16	5	2

(a) Why does the yield of ammonia decrease with increased temperature? (2)

(b) Why does the yield of ammonia increase with increased pressure? (2)

(c) Why are conditions of 200 atm pressure and 450°C used in the industrial process? (3)

(d) Suggest a better way than a table to present this data. (1)

[Higher]

4 The reaction to produce poly(ethene) is exothermic.

$$n\ C_2H_4 \rightarrow \text{--}(CH_2\text{---}CH_2)_n\text{--}$$
ethene   poly(ethene)

The conditions used in two processes to make poly(ethene) are shown in the table.

Process	Temperature (°C)	Pressure (atm)	Catalyst
A	150–300	1 000–3000	no
B	40–80	1–50	yes

(a) What enables process **B** to be operated under less vigorous conditions? (1)

(b) Suggest one way to keep the energy used to a minimum in both processes. (1)

(c) Suggest **two** environmental advantages of using process **B** to make poly(ethene). (2)

5 A student had learned that the reaction between hydrochloric acid and sodium hydroxide solution was exothermic. She, therefore, predicted that when she added more acid to the alkali more heat would be produced. She used a burette to deliver exact amounts of hydrochloric acid to 20 cm³ of alkali in a flask. She used a thermometer to measure the temperature. Her results are in this table:

Volume of acid added (cm³)	Temperature recorded (°C)
0	17
10	21
20	24
30	21
40	21
50	20

(a) How should she have insulated the flask? (1)

(b) Explain why she should have taken the temperature of the acid before adding it to the sodium hydroxide solution. (2)

(c) Did she actually measure the heat produced by the reaction? Explain your answer. (1)

(d) How might she have used an indicator to increase the accuracy of her method? (1)

HOW SCIENCE WORKS QUESTIONS

Jack set up some apparatus to see the effect of temperature on the rate of a reaction between calcium carbonate and hydrochloric acid.

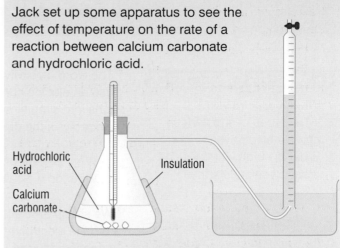

Jack was careful to ensure that the mass of the calcium carbonate, the concentration and the volume of the hydrochloric acid were kept the same for the start of each experiment. He also ensured that the temperature of the reactants was checked after the carbon dioxide had been collected. He timed how long it took, at each temperature, to fill the burette.

Here are his results:

Temperature of reactants (°C)	Average time taken (s)	Average temperature change (°C)
15	145	+1
20	105	+1
25	73	+3
30	51	+3
35	30	+4

a) Plot a graph of the temperature of the reactants against the average time taken. (3)

b) Describe the pattern that you think is shown by these results. (2)

c) List three variables that Jack controlled in this investigation. (3)

d) Name a variable that Jack could not control, but did take account of. (1)

e) What type of error was the changing temperature? Explain your answer. (1)

f) Was the sensitivity of the thermometer good enough? Explain your answer. (1)

g) Why is it not possible to judge the precision of Jack's results? (2)

h) Do you doubt the reliability of Jack's results? Explain your answer. (1)

C2 6.1   Electrolysis – the basics

Figure 1 The first person to explain electrolysis was Michael Faraday, who worked on this and many other problems in science nearly 200 years ago. His work formed the basis of an understanding of electrolysis that we still use today.

The word electrolysis means 'splitting up using electricity'. In electrolysis we use an electric current to break down (or **decompose**) a substance made of ions into simpler substances. We call the substance broken down by electrolysis the **electrolyte**.

a) What is electrolysis?
b) What do we call the substance broken down by electrolysis?

We set up an electrical circuit for electrolysis that has two electrodes which dip into the electrolyte. The electrodes are conducting rods. One of these is connected to the positive terminal of a power supply, the other is connected to the negative terminal.

We normally make the electrodes out of an unreactive (or **inert**) substance like graphite or platinum. This is so they do not react with either the electrolyte or the products made during electrolysis. We use the name **anode** for positive electrode, while we call the negative electrode the **cathode**.

During electrolysis, positively charged ions move to the negative electrode (cathode) and negative ions move to the positive electrode (anode).

When the ions reach the electrodes they can lose their charge and be deposited as elements. Depending on the compound being electrolysed, gases may be given off or metals deposited at the electrodes.

DEMONSTRATION

The electrolysis of lead bromide

This demonstration needs a fume cupboard because bromine is toxic and corrosive.

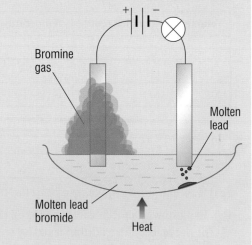

Bromine gas

Molten lead

Molten lead bromide

Heat

Figure 2 When we pass electricity through molten lead bromide it forms molten lead and brown bromine gas as the electrolyte is broken down by the electricity

● When does the bulb light up?

Figure 2 shows how electricity breaks down lead bromide into lead and bromine:

$$\text{lead bromide} \rightarrow \text{lead} + \text{bromine}$$
$$PbBr_2\ (l) \rightarrow Pb\ (l) + Br_2\ (g)$$

Lead bromide is an ionic substance which does not conduct electricity when it is solid. But when we melt it the ions can move freely towards the electrodes.

The positive lead ions move towards the cathode, while the negatively charged bromide ions move towards the anode. Notice how the state symbols in the equation tell us that the lead bromide and the lead are molten. The '(l)' stands for 'liquid', while bromine is given off as a gas, shown as '(g)'.

c) Which electrode do positive ions move towards during electrolysis?
d) Which electrode do negative ions move towards during electrolysis?

Many ionic substances have very high melting points. This can make electrolysis very difficult or even impossible. But some ionic substances dissolve in water, and when this happens the ions can move freely.

When we dissolve ionic substances in water to electrolyse them it is more difficult to predict what will be formed. This is because water also forms ions, and so the product at the anode and the cathode is not always exactly what we expect.

When we electrolyse a solution of copper bromide in water, copper ions move to the negative electrode (cathode) and the bromide ions move to the positive electrode (anode). Copper bromide is split into its two elements at the electrodes:

$$\text{copper bromide} \rightarrow \text{copper} + \text{bromine}$$
$$CuBr_2 \text{ (aq)} \quad \rightarrow \quad Cu \text{ (s)} + Br_2 \text{ (aq)}$$

In this case the state symbols in the equation tell us that the copper bromide is dissolved in water, shown as '(aq)'. The elements that are produced are solid copper, shown as '(s)', and bromine which remains dissolved in the water – '(aq)'.

Covalent compounds cannot be split by electrolysis.

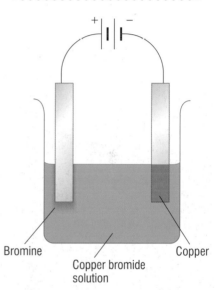

Bromine | Copper
Copper bromide solution

Figure 3 If we dissolve copper bromide in water we can decompose it by electrolysis. Copper metal is formed at the cathode, while brown bromine appears in solution around the anode.

SUMMARY QUESTIONS

1 Copy and complete using the words below:

> anode   cathode   ions   molten   move
> negative   solution

In electrolysis the …… is the positive electrode while the …… is the …… electrode. For the current to flow, the …… must be able to …… between the electrodes. This can only happen if the substance is in …… or if it is …… .

2 Predict the products formed at each electrode when the following compounds are melted and then electrolysed:

a) copper iodide
b) potassium bromide
c) sodium fluoride.

3 Solid ionic substances do not conduct electricity. Using words and diagrams explain why they conduct electricity when molten or in solution.

KEY POINTS

1 Electrolysis involves splitting up a substance using electricity.
2 Ionic substances can be electrolysed when they are molten or in solution.
3 In electrolysis positive ions move to the negative electrode (cathode) and negative ions move to the positive electrode (anode).

C2 6.2

Changes at the electrodes

LEARNING OBJECTIVES

1 What happens during electrolysis?
2 How can we represent what happens in electrolysis? [Higher]
3 How does water affect the products of electrolysis?

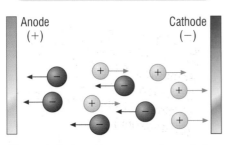

Figure 1 An ion always moves towards the oppositely charged electrode

During electrolysis ions move towards the electrodes. The direction they move in depends on their charge. As we saw on the previous page, positive ions move towards the negative electrode (the cathode). Negative ions move towards the positive electrode (the anode).

When ions reach an electrode, they either lose or gain electrons depending on their charge.

Negatively charged ions **lose** electrons to become neutral atoms. Positively charged ions form neutral atoms by **gaining** electrons.

a) How do negatively charged ions become neutral atoms in electrolysis?
b) How do positively charged ions become neutral atoms in electrolysis?

The easiest way to think about this is to look at an example:

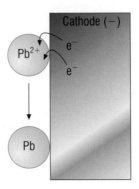

In the electrolysis of molten lead bromide, positively charged lead ions (Pb^{2+}) move towards the cathode ($-$). When they get there, each ion gains **two** electrons to become a neutral lead atom.

Gaining electrons is called **reduction** – we say that the lead ions are **reduced**. 'Reduction' is simply another way of saying 'gaining electrons'.

When molten lead bromide is electrolysed, negatively charged bromide ions (Br^-) move towards the anode ($+$). When they get there, each ion loses **one** electron to become a neutral bromine atom. Two bromine atoms then form a covalent bond to make a bromine molecule, Br_2.

Losing electrons is called **oxidation** – we say that the bromide ions are **oxidised**. 'Oxidation' is another way of saying 'losing electrons'.

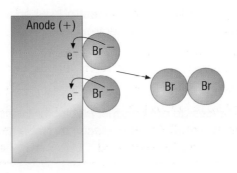

We represent what is happening at the electrodes using **half equations**. We call them this because what happens at one electrode is only half the story – we need to know what is happening at both electrodes to know what is happening in the whole reaction.

At the negative electrode:

$$Pb^{2+} + 2e^- \rightarrow Pb \qquad \text{(notice how an electron is written as 'e}^-\text{')}$$

At the positive electrode:

$$2Br^- \rightarrow Br_2 + 2e^-$$

Sometimes half equations are written showing the electrons being removed from negative ions, like this:

$$2Br^- - 2e^- \rightarrow Br_2$$

Neither method is more 'right' than the other – it just depends on how you want to write the half equation.

Because **RED**uction and **OX**idation take place at the same time in electrolysis (reduction at the cathode (−), oxidation at the anode (+)), it is sometimes called a **redox** reaction.

The effect of water

When we carry out electrolysis in water the situation is made more complicated by the fact that water contains ions. The rule for working out what will happen is to remember that if two elements can be produced at an electrode, the less reactive element will usually be formed.

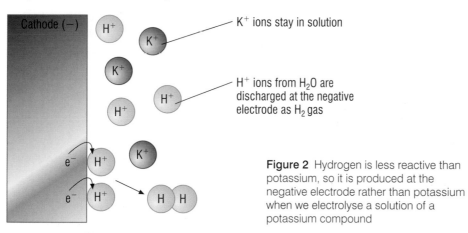

K⁺ ions stay in solution

H⁺ ions from H_2O are discharged at the negative electrode as H_2 gas

Figure 2 Hydrogen is less reactive than potassium, so it is produced at the negative electrode rather than potassium when we electrolyse a solution of a potassium compound

GET IT RIGHT!

Remember OILRIG – Oxidation Is Loss (of electrons), Reduction Is Gain (of electrons).

SUMMARY QUESTIONS

1 Copy and complete using the words below:

anode (+)   cathode (−)   electrodes   gain   less   lose   ions   oxidised   reduced

During electrolysis ...... move towards the .... At the ...... positively charged ions are ...... and ...... electrons. At the ...... negatively charged ions are ...... and ...... electrons. When electrolysis is carried out in water, the ...... reactive element is usually produced.

2 Copy and complete the following half-equations where necessary:

a) $Cl^- \rightarrow Cl_2 + e^-$ c) $Ca^{2+} + e^- \rightarrow Ca$ e) $Na^+ + e^- \rightarrow Na$

b) $O^{2-} \rightarrow O_2 + e^-$ d) $Al^{3+} + e^- \rightarrow Al$ f) $H^+ + e^- \rightarrow H_2$ [Higher]

KEY POINTS

1 In electrolysis, the ions move towards the oppositely charged electrodes.

2 At the electrodes, negative ions are oxidised while positive ions are reduced.

3 Reactions where reduction and oxidation happen are called redox reactions.

4 When electrolysis happens in water, the less reactive element is usually produced at an electrode.

C2 6.3

Electrolysing brine

The electrolysis of brine (sodium chloride solution) is an enormously important industrial process. When we pass an electric current through brine we get three products. Chlorine gas is produced at the positive electrode, hydrogen gas is made at the negative electrode, and a solution of sodium hydroxide is also formed:

sodium chloride solution $\xrightarrow{\text{electrolysis}}$ hydrogen + chlorine + sodium hydroxide solution

a) What are the three products made when we electrolyse brine?

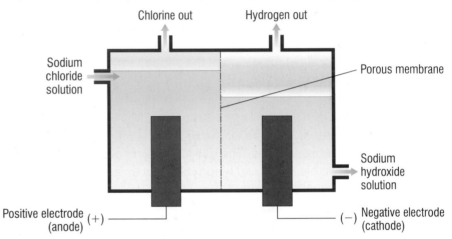

Figure 1 Brine can be electrolysed in a cell in which the two electrodes are separated by a porous membrane. This is called a **diaphragm cell**.

The half equations for what happens in the electrolysis of brine are:

At the positive electrode, $2Cl^- (aq) \rightarrow Cl_2 (g) + 2e^-$

At the negative electrode, $2H^+ (aq) + 2e^- \rightarrow H_2 (g)$

This leaves a solution containing Na^+ and OH^- ions, i.e. a solution of sodium hydroxide.

HIGHER

Using chlorine

Chlorine is a poisonous green gas which causes great damage to our bodies if it is inhaled in even tiny quantities. But it is also a tremendously useful chemical. The chlorine made when we electrolyse brine plays a vital role in public health. It is used to kill bacteria in drinking water and in swimming pools.

We can also react chlorine with the sodium hydroxide produced in the electrolysis of brine. This makes a solution called **bleach** (sodium chlorate(I)). Bleach is a strong oxidising agent which is very good at killing bacteria. We use it widely in homes, hospitals and industry to maintain good hygiene.

Chlorine is also an important part of many other disinfectants as well as the plastic (polymer) known as PVC.

b) What is chlorine used for?

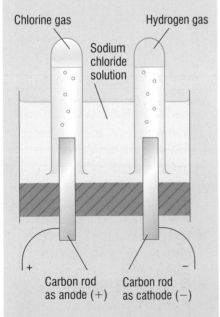

Using hydrogen

The hydrogen that we make by electrolysing brine is particularly pure. This makes it very useful in the food industry. We make margarine by reacting hydrogen with vegetable oils under pressure and with a catalyst to turn the oil into a soft spreadable solid.

We can also react hydrogen with the chlorine made by the electrolysis of brine to make hydrogen chloride gas. We can then dissolve this gas in water to make hydrochloric acid. This very pure acid is used widely by the food and pharmaceutical industries.

c) What is hydrogen used for?

Using sodium hydroxide

The sodium hydroxide which is made when we electrolyse brine is used to make soap and paper. It is also used to control the pH in many industrial processes. The other major use of sodium hydroxide is to combine it with the chlorine produced to make bleach (see previous page).

d) What is sodium hydroxide used for?

Figure 2 Chlorine brings us clean, disease-free drinking water and helps to keep our homes, schools and hospitals free from disease

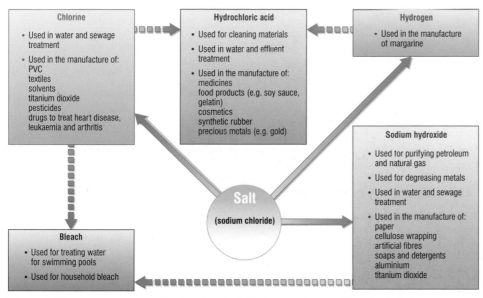

Figure 3 There are many uses for the chemicals that we get from salt by electrolysis

KEY POINTS

1 When we electrolyse brine we get three products – chlorine gas, hydrogen gas and sodium hydroxide solution.
2 Chlorine is used to kill microbes in drinking water and swimming pools, and to make hydrochloric acid, disinfectants, bleach and plastics.
3 Hydrogen is used to make margarine and hydrochloric acid.
4 Sodium hydroxide is used to make bleach, paper and soap.

SUMMARY QUESTIONS

1 Copy and complete using the words below:

   **chlorine   hydrochloric   hydrogen   sodium chlorate(I)
   sodium hydroxide**

   When we pass an electric current through brine we make ...... gas, ... gas and .... solution. These products are also used to make .... solution (bleach) and ...... acid.

2 a) Write a balanced chemical equation to show the production of chlorine, hydrogen and sodium hydroxide from salt solution by electrolysis. The equation is started off for you below:

   $$NaCl + H_2O \rightarrow ?$$ [Higher]

   b) We can also electrolyse *molten* sodium chloride. Compare the products formed with those from the electrolysis of sodium chloride solution. What are the differences?

C2 6.4 Purifying copper

LEARNING OBJECTIVES

1 Why do we need to purify copper?
2 How do we use electrolysis to purify copper?

When we remove copper from its ore it is possible to get copper that is about 99% pure. The impurities include precious metals like gold, silver and platinum. These affect the conductivity of the copper, and must be removed before we can use the copper for electrical wires.

a) What impurities may be found in copper after it has been removed from its ore?
b) Why must these be removed?

Figure 1 A major use of copper is to make cables and wires for carrying electricity and electrical signals

We purify copper using electrolysis. A bar of impure copper is used as the anode (+), and a thin sheet of pure copper is the cathode (−). The electrolysis takes place in a solution containing copper ions (usually copper sulfate solution).

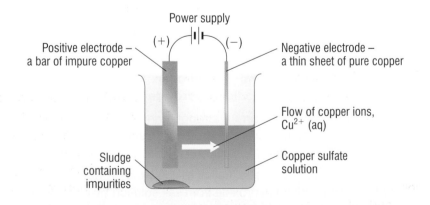

Power supply

Positive electrode – a bar of impure copper

(+)   (−)

Negative electrode – a thin sheet of pure copper

Flow of copper ions, Cu^{2+} (aq)

Sludge containing impurities

Copper sulfate solution

Figure 2 Copper is refined using electrolysis

At the positive electrode, copper atoms are oxidised. They form copper ions and go into the solution:

$$Cu \ (s) \rightarrow Cu^{2+} \ (aq) + 2 e^-$$

[Higher]

At the negative electrode, copper ions are reduced. They form copper atoms which are deposited on the electrode:

$$Cu^{2+} (aq) + 2e^- \rightarrow Cu (s) \qquad \text{[Higher]}$$

c) Where are the copper atoms oxidised?

d) What is formed when copper atoms are oxidised?

Once we have purified the copper in the electrolytic cell, it is removed, melted and then formed into bars or ingots.

The sludge, containing precious metal impurities, is periodically removed from the electrolysis cell to collect the precious metals from it.

PRACTICAL

Comparing electrodes

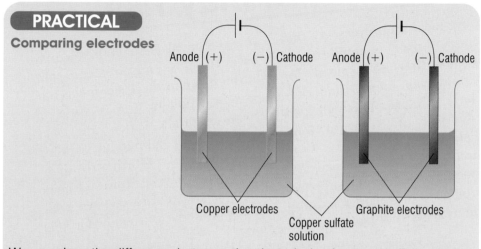

Anode (+)　　(−) Cathode　　Anode (+)　　(−) Cathode

Copper electrodes　　　Graphite electrodes

Copper sulfate solution

We can show the difference between the electrolysis of copper sulfate solution using copper electrodes and electrolysis using graphite electrodes.

- What happens at each electrode?

Here is a summary of the electrolysis of copper sulfate solution using different electrodes:

HIGHER

Using copper electrodes		Using graphite electrodes	
At anode (+)	**At cathode (−)**	**At anode (+)**	**At cathode (−)**
$Cu(s) \rightarrow Cu^{2+}(aq)$ $+ 2e^-$	$Cu^{2+}(aq) + 2e^-$ $\rightarrow Cu(s)$	$2H_2O(l) \rightarrow 4H^+(aq)$ $+ O_2(g) + 4e^-$	$Cu^{2+}(aq) + 2e^-$ $\rightarrow Cu(s)$

We can also show the half equation at the graphite anode (+) as:

$$4OH^-(aq) \rightarrow 2H_2O(l) + O_2(g) + 4e^-$$

SUMMARY QUESTIONS

1 Copy and complete using the words below:

　　atoms　　cathode (−)　　copper　　copper sulfate　　deposited
　　electrolysis　　electrons　　impure　　oxidised　　reduced

　　Copper is purified by ...... using electrodes made of .... An electric current is passed through a solution of .... The anode (+) is made of ...... copper. The copper atoms are ...... and go into the solution. At the ...... they gain ...... and are .... They form copper ...... and are ...... on the cathode (−).

2 What happens to the impurities that are removed from the copper when it is purified?

KEY POINTS

1 Copper extracted from its ore contains impurities such as gold and silver.

2 Copper is purified by electrolysis to remove these impurities.

To build or not to build?

Unemployment in Newtown may rise!

Two big local employers say that concerns over supplies of chemicals that they need for their factories mean that they may have to close. This will lead to hundreds of Newtown jobs being slashed.

A director of Allied Fats said 'We have been worried about supplies of hydrogen to our plant for some time since the cost of transporting this chemical is so high. We may have to close and relocate our business somewhere nearer to our present suppliers.'

Consolidated Paper are also worried about supplies of sodium hydroxide and chlorine to their paper mill in the town. Tracey Wiggins, the MP for Newtown, said 'This would be a tragedy for the town.'

Hope for new employment!

Following concerns about supplies of chemicals to two big local employers, we can exclusively reveal that a deal is being struck that would bring a manufacturer of these chemicals to Newtown.

BrineCo, one of the largest chemical companies in the country, is currently in talks with the council about building a big new plant to produce chemicals in a new factory near the town. BrineCo already manufacture chlorine and sodium hydroxide at other plants in the UK.

Local MP Tracey Wiggins said 'This would be a wonderful opportunity for workers and their families in Newtown and the surrounding area.'

QUESTIONS

Look at the two leaflets produced by BrineCo and by the local pressure group GREEN.

1 Make a list of the differences between the two maps on the leaflets.
2 How is BrineCo trying to persuade people that their factory is a good idea?
3 How is GREEN trying to persuade people that the factory is *not* a good idea?

ACTIVITY

Write an editorial for the local newspaper in which you examine both sides of the argument for bringing the BrineCo chemical factory to the town. The final part of your editorial should come down on one side or the other – but you must argue your point logically. You may also decide that the factory should go ahead, but on a different site. Can *you* persuade local people that *you* are the voice of reason?

BrineCo
working for you!

BrineCo produce chlorine and sodium hydroxide solution by passing electricity through brine (salt solution). This is called **electrolysis**.

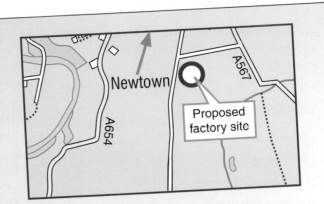

The chlorine that we make is used to make paper, chemicals and plastics, and for treating water to kill bacteria. Sodium hydroxide solution is sold to companies making paper, artificial fibres, soaps and detergents.

Our new factory can bring many benefits to Newtown. Consolidated Paper will be a major user of BrineCo's chemicals, and Allied Fats will buy the hydrogen produced by our factory.

This will give both companies a cheaper supply of raw materials than they have at present. *Think carefully about BrineCo's proposals – they mean a secure future for you and your children.*

KEEP NEWTOWN FREE FROM CHEMICALS!!!

GO GREEN
Keep Newtown clean!

Give
Rights to
Everyone's
Environment in
Newtown

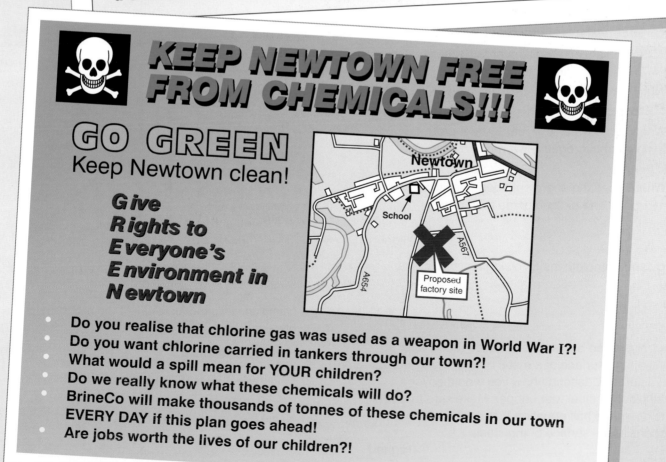

- Do you realise that chlorine gas was used as a weapon in World War I?!
- Do you want chlorine carried in tankers through our town?!
- What would a spill mean for YOUR children?
- Do we really know what these chemicals will do?
- BrineCo will make thousands of tonnes of these chemicals in our town EVERY DAY if this plan goes ahead!
- Are jobs worth the lives of our children?!

SUMMARY QUESTIONS

1 Select A or B to describe correctly what happens at the positive electrode and negative electrode in electrolysis for a) to f).

a) Positive ions move towards this.   A Positive electrode

b) Negative ions move towards this.   B Negative electrode

c) Reduction happens here.

d) Oxidation happens here.

e) Connected to the negative terminal of the power supply.

f) Connected to the positive terminal of the power supply.

2 Make a table to show which of the following ions would move towards the positive electrode and which towards the negative electrode during electrolysis. (You may need to use a copy of the periodic table to help you.)

sodium, iodide, calcium, fluoride, zinc, oxide, aluminium, bromide

3 Water can be split into hydrogen and oxygen using electrolysis. The word equation for this reaction is:

water → hydrogen + oxygen

a) Write a balanced equation for this reaction using the correct chemical symbols.

b) Write half-equations to show what happens at the positive and negative electrodes.

c) When some water is electrolysed it produces 2 moles of hydrogen. How much oxygen is produced?

d) Where does the energy needed to split water into hydrogen and oxygen come from during electrolysis?

[Higher]

4 Copy and complete the following half equations:

a) $K^+ \rightarrow K$ b) $Ba^{2+} \rightarrow Ba$

c) $I^- \rightarrow I_2$ d) $O^{2-} \rightarrow O_2$

[Higher]

5 Electrolysis can be used to produce a thin layer of metal on the surface of a metal object. Using words and diagrams, describe how you would cover a small piece of steel with copper. Make sure that you write down the half equation that describes what happens at the surface of the steel.

[Higher]

EXAM-STYLE QUESTIONS

1 The table shows the results of passing electricity through some substances. Carbon electrodes were used.

Substance	Product at negative electrode	Product at positive electrode
Molten lead bromide	lead	A
Molten B	magnesium	chlorine
Aqueous sodium sulfate solution	C	oxygen
Aqueous copper sulfate solution	D	E

(a) Name A, B, C, D and E. (5)

(b) What is the name used for substances that conduct electricity and are decomposed by it? (1)

(c) Why must the substances be molten or in solution? (1)

(d) Explain why reduction takes place at the negative electrode. (2)

2 The diagram shows a cell used for the electrolysis of brine. Brine is a solution of sodium chloride in water.

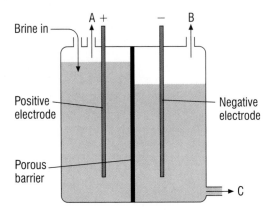

(a) Name and give the formulae of the positive ions in brine. (2)

(b) Name and give the formulae of the negative ions in brine. (2)

(c) Name gases A and B. (2)

(d) Explain as fully as you can how gas B is produced. (4)

(e) Name the product in solution C. (1)

3 Mild steel can be electroplated with tin in the laboratory. The diagram shows the apparatus used.

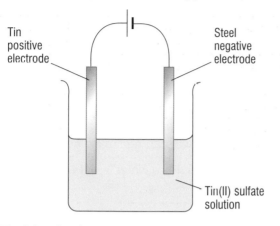

Tin positive electrode

Steel negative electrode

Tin(II) sulfate solution

(a) Explain what happens at the negative electrode to deposit tin on the steel. (3)

(b) What happens to tin at the positive electrode? (3)

(c) Why does the concentration of the tin(II) sulfate solution not change during the electrolysis? (2)

(d) Some food cans are made of mild steel coated with tin. Suggest **two** reasons why tin plated steel is chosen for this use. (2)

4 A student was interested in electrolysis. He knew that a current passes through a copper sulfate solution. With copper electrodes some of the copper would come away from one electrode and move to the other. He thought that there would be the same amount of copper leaving one electrode as attached to the other electrode. He set up his equipment and weighed each electrode several times over a 25 minute period. His results are in this table:

Loss in mass of positive electrode in 5 minutes (g)	Gain in mass of negative electrode in 5 minutes (g)	Time (mins)
0.027	0.021	5
0.022	0.027	10
0.061	0.030	15
0.001	0.025	20
0.025	0.027	25

(a) What evidence is there for an anomalous result? (1)

(b) What was the range for the results at the negative electrode? (1)

(c) What evidence is there to support the student's prediction? (1)

(d) Comment on the reliability of the results. (1)

HOW SCIENCE WORKS QUESTIONS

Hydrogen – the new petrol?

Hydrogen could be a very important fuel for personal transport. However, there are many practical problems to be solved.

Mikael came up with the idea of using the Sun's energy to produce hydrogen from sea water. The apparatus used was similar to what you might have seen in your school laboratory. However, he used a solar cell to produce a voltage to drive the electrolysis. Mikael left the electrolysis for the same time for each solution used.

The results are shown below.

Solution used	Volume of hydrogen (cm³)			mean
Sea water	33	27	45	

a) Calculate the mean volume of hydrogen produced. (1)

b) What do these results tell us about the precision of the method used? Explain your answer. (1)

c) What probably caused the variation in the student's results? (1)

d) Mikael's teacher dismissed the research saying, 'It could never come to anything that might produce large volumes of hydrogen.'
Why do you think the teacher thought this? (1)

e) Mikael's dad thought it was a brilliant idea and a chance to make some money! He pictured a huge factory turning out millions of tonnes of hydrogen and millions of pounds of money! He quizzed Mikael about his results. He asked Mikael if he was telling the whole truth. Why was it important that Mikael was telling the whole truth about his investigation? (1)

f) What might be Mikael's next step towards becoming a millionaire? (1)

C2 7.1 Acids and alkalis

ACIDS, ALKALIS AND SALTS

LEARNING OBJECTIVES

1 Why are solutions acidic or alkaline?
2 How do we measure acidity?

Figure 1 Acids and bases are all around us, in many of the things we buy at the shops, in our schools and factories – and inside us too

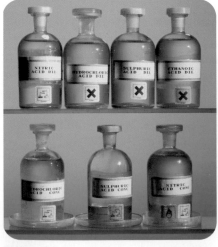

Figure 2 Some common laboratory acids

Acids and bases are an important part of our understanding of chemistry. They play a vital part inside us and for all other living things too.

What are acids and bases?

When we dissolve a substance in water we make an **aqueous solution**. The solution may be acidic, alkaline or neutral, depending on the chemical we have dissolved. **Bases** are chemicals which can neutralise **acids**.

Alkalis are bases which dissolve in water. Pure water is **neutral**.

a) What is a base?
b) What is an alkali?

Acids include chemicals like citric acid, sulfuric acid and ethanoic acid. All acids taste very sour, although many acids are far too dangerous to put in your mouth. We use acids in many chemical reactions in the laboratory. Ethanoic acid (vinegar) and citric acid (the sour taste in citric fruit, fizzy drinks and squashes) are acids which we regularly eat.

One acid that we use in the laboratory is hydrochloric acid. This is formed when the gas hydrogen chloride (HCl) dissolves in water:

$$HCl\ (g) \xrightarrow{\text{water}} H^+\ (aq) + Cl^-\ (aq)$$

All acids form H^+ ions when we add them to water – it is hydrogen ions that make a solution acidic. Hydrogen chloride also forms chloride ions (Cl^-). The '(aq)' symbol shows that the ions are in an 'aqueous solution'. In other words, they are dissolved in water.

c) What ions do all acids form when we add them to water?

Bases are the opposite of acids in the way they react. Because alkalis are bases which dissolve in water they are the bases which we use most commonly. For example, sodium hydroxide solution is often found in our school laboratories. Sodium hydroxide solution is formed when we dissolve solid sodium hydroxide in water:

$$NaOH\ (s) \xrightarrow{\text{water}} Na^+\ (aq) + OH^-\ (aq)$$

All alkalis form hydroxide ions (OH^-) when we add them to water. It is hydroxide ions that make a solution alkaline.

d) What ions do all alkalis form when we add them to water?

Measuring acidity

Indicators are special chemicals which change colour when we add them to acids and alkalis. Litmus paper is one well-known indicator, but there are many more. These include some natural ones like the juice of red cabbage or beetroot.

We use the **pH scale** to show how acid or alkaline a solution is. The scale runs from 0 (most acidic) to 14 (most alkaline). **Universal indicator** is a very special indicator made from a number of dyes. It turns different colours at different values of pH. Anything in the middle of the pH scale (pH 7) is **neutral**, neither acid nor alkali.

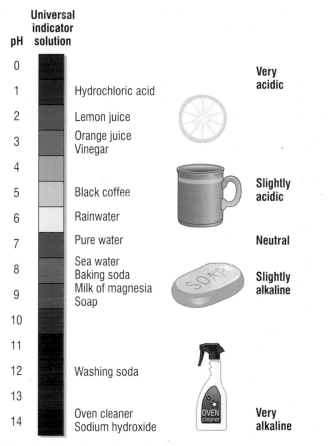

Figure 3 The pH scale tells us how acid or alkaline a solution is

A H$^+$ ion is simply a hydrogen atom that has lost an electron – in other words a proton. So another way of describing an acid is to say that it is a 'proton donor'.

PRACTICAL

Which is the most alkaline product?

Compare the alkalinity of various cleaning products.

You can test washing-up liquids, shampoos, soaps, hand-washing liquids, washing powders/liquids and dishwasher powders/tablets.

You might be able to use a pH sensor and data logger to collect your data.

● What are the advantages of using a pH sensor instead of universal indicator solution or paper?

SUMMARY QUESTIONS

1 Copy and complete using the words below:

   alkaline   dissolve   greater   hydrogen   hydroxide   less   neutralise   pH   seven

   Acids form ...... ions when we dissolve them in water. Bases react with acids and ...... them. Alkalis are bases which ...... in water. They form ...... ions when they do this. The ...... scale tells us how acidic or alkaline a solution is. If the pH is ...... the solution is neutral, if it is ...... than 7 the solution is acidic, and if it is ...... than 7 the solution is ....

2 How could you use paper containing universal indicator as a way of distinguishing between pure water, sodium hydroxide solution and citric acid solution?

KEY POINTS

1 Acids are substances which produce H$^+$ ions when we add them to water.
2 Bases are substances that will neutralise acids.
3 An alkali is a soluble base. Alkalis produce OH$^-$ ions when we add them to water.
4 We use the pH scale to show how acidic or alkaline a solution is.

C2 7.2   Making salts from metals or bases

1 What do we make when we react acids and metals?
2 What do we make when we react acids and bases?
3 What salts can we make?

Acids + metals

We can make salts by reacting acids with metals. This is only possible if the metal is above hydrogen in the reactivity series. If it is, then hydrogen gas is produced when the acid reacts with the metal, and a salt is also produced:

$$acid \quad + \quad metal \quad \rightarrow \quad salt \quad + \; hydrogen$$

$$2HCl\,(aq) \quad + \quad Mg\,(s) \quad \rightarrow \quad MgCl_2\,(aq) \quad + \quad H_2\,(g)$$

hydrochloric acid + magnesium → magnesium chloride + hydrogen
solution

a) What does the reaction between an acid and a metal produce?

Acid + insoluble base

When we react an acid with a base we produce a solution which contains a salt and water.

The general equation which describes all reactions of this type is:

$$acid + base \rightarrow salt + water$$

b) What two substances are formed when an acid and a base react?

The salt that we make depends on the metal or the base that we use in the reaction and the acid. So bases that contain sodium ions will always make sodium salts, while those that contain potassium ions will always make potassium salts.

In a similar way:

- the salts formed when we neutralise hydrochloric acid are always *chlorides*
- sulfuric acid always makes salts which are *sulfates*, and
- nitric acid always makes *nitrates*.

The oxide of a transition metal, such as iron(III) oxide, is an example of a base that we can use to make a salt in this way:

$$acid \quad + \quad base \quad \rightarrow \quad salt \quad + \quad water$$

$$6HCl\,(aq) \quad + \quad Fe_2O_3\,(s) \quad \rightarrow \quad 2FeCl_3\,(aq) \quad + \; 3H_2O\,(l)$$

hydrochloric acid + solid iron(III) oxide → iron(III) chloride + water
solution

c) Name the salt formed when dilute sulfuric acid reacts with zinc oxide.

PRACTICAL

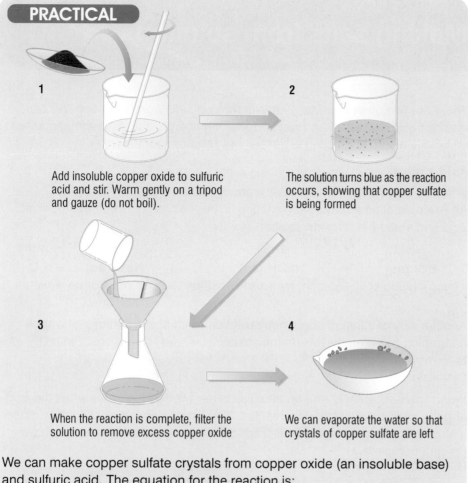

1 Add insoluble copper oxide to sulfuric acid and stir. Warm gently on a tripod and gauze (do not boil).

2 The solution turns blue as the reaction occurs, showing that copper sulfate is being formed

3 When the reaction is complete, filter the solution to remove excess copper oxide

4 We can evaporate the water so that crystals of copper sulfate are left

We can make copper sulfate crystals from copper oxide (an insoluble base) and sulfuric acid. The equation for the reaction is:

acid   +   base   →   salt   + water

$$H_2SO_4 (aq) + \quad CuO (s) \quad → \quad CuSO_4 (aq) + H_2O (l)$$

sulfuric   + solid copper(II) → copper sulfate + water
acid   oxide   solution

- What does the copper sulfate look like? Draw a diagram if necessary.

SUMMARY QUESTIONS

1 Copy and complete using the words below:

bases   hydrogen   metals   neutralisation   salt   water

The reaction between an acid and a base is called a …… reaction. When this happens, a …… is formed, together with …… . Salts can be made by reacting acids with ……, when …… gas is formed along with the salt. They can also be made by reacting acids with insoluble ……, when water is formed as well as the salt.

2 'Bicarbonate for bees and vinegar for vasps (wasps!!)' is one way of remembering what to do if you are stung by a bee or a wasp. What does this suggest about the pH of bee stings and wasp stings?

C2 7.3   Making salts from solutions

LEARNING OBJECTIVES

1 How can we make salts from an acid and an alkali?
2 How can we make insoluble salts?
3 How can we remove unwanted ions from solutions?

Figure 2 Ammonium nitrate is used as a fertiliser

Figure 3 Water treatment plants use chemical treatments to precipitate chemical compounds that can then be removed by filtering the solution

There are two other important ways of making salts from solutions. We can react an acid and an alkali together to form a soluble salt. And sometimes we can make a salt by reacting two other salt solutions together.

Acid + alkali

When an acid reacts with an alkali a neutralisation reaction takes place. An example of a neutralisation reaction is the reaction between hydrochloric acid and sodium hydroxide solution:

$$\text{acid} + \text{alkali} \rightarrow \text{salt} + \text{water}$$

$$\text{HCl (aq)} + \text{NaOH (aq)} \rightarrow \text{NaCl (aq)} + \text{H}_2\text{O (l)}$$

hydrochloric acid + sodium hydroxide solution → sodium chloride + water solution

Another way of thinking about neutralisation reactions is in terms of what is happening between the ions in the solutions, where H^+ ions react with OH^- ions to form water:

$$H^+ \text{(aq)} + OH^- \text{(aq)} \rightarrow H_2O \text{(l)}$$

When we react an acid and an **alkali** together we need to know when the acid and the alkali have completely reacted. We can use an indicator for this, since a strong acid and a strong alkali produce a neutral solution when they have reacted completely.

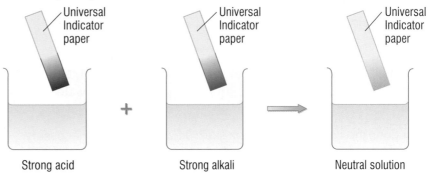

Strong acid   Strong alkali   Neutral solution

Figure 1 Universal indicator paper can show us when a strong acid and a strong alkali have reacted completely to form a salt because a neutral solution is formed

We can make ammonium salts by reacting an acid and alkali together. Ammonia reacts with water to form ammonium hydroxide (a weak alkali):

$$NH_3 \text{(aq)} + H_2O \text{(l)} \rightleftharpoons NH_4OH \text{(aq)}$$

Ammonium hydroxide then reacts with an acid (for example, nitric acid):

$$\text{acid} + \text{alkali} \rightarrow \text{salt} + \text{water}$$

$$\text{HNO}_3 \text{(aq)} + \text{NH}_4\text{OH (aq)} \rightarrow \text{NH}_4\text{NO}_3 \text{(aq)} + \text{H}_2\text{O (l)}$$

nitric acid + ammonium hydroxide → ammonium nitrate + water solution   solution

Ammonium nitrate contains a large amount of nitrogen, and it is very soluble in water. This makes it ideal as a source of nitrogen to replace the nitrogen taken up from the soil by plants as they grow.

a) Write down a general equation for the reaction between an acid and an alkali.

Making insoluble salts

We can sometimes make salts by combining two solutions. When this makes an insoluble salt, we call the reaction a **precipitation** reaction because the insoluble solid that is formed is called a **precipitate**.

$$Pb(NO_3)_2 \text{ (aq)} + 2\,NaCl \text{ (aq)} \rightarrow PbCl_2 \text{ (s)} + 2\,NaNO_3 \text{ (aq)}$$

| lead nitrate | + sodium chloride | → solid lead chloride | + sodium nitrate |
| solution | solution | (precipitate) | solution |

The equation for the reaction shows how the lead chloride that forms is insoluble in water. It forms a solid precipitate that we can filter off from the solution.

b) What do we call a reaction that produces a precipitate?

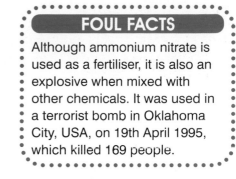
PRACTICAL

Making an insoluble salt

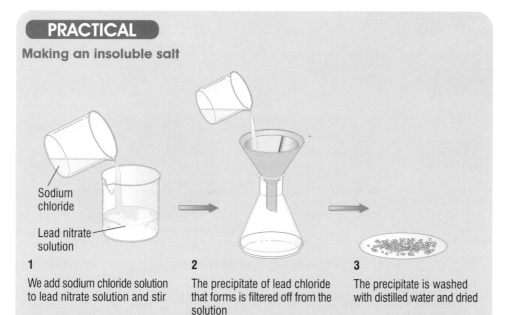

Sodium chloride

Lead nitrate solution

1 We add sodium chloride solution to lead nitrate solution and stir

2 The precipitate of lead chloride that forms is filtered off from the solution

3 The precipitate is washed with distilled water and dried

We can make lead chloride crystals from lead nitrate solution and sodium chloride solution. The equation for the reaction is shown at the top of this page.

● What does the lead chloride look like?

Using precipitation

We use precipitation reactions to remove pollutants from the wastewater from factories and industrial parks before the effluent is discharged into rivers and the sea.

An important precipitation reaction is the removal of metal ions from water that has been used in industrial processes. By raising the pH of the water, we can make insoluble metal hydroxides precipitate out of the solution. This produces a sludge which we can easily remove from the solution.

The cleaned-up water can then be discharged safely into a river or into the sea.

SUMMARY QUESTIONS

1 Copy and complete using the words below:

   acid   alkali   insoluble   metal   polluted   precipitation
            solid   soluble   water

   We can make salts by reacting an …… with an …… . This makes the salt
   and …… . We can also make …… salts by reacting two …… salts
   together. We call this a …… reaction because the salt is formed as a
   …… . This type of reaction is also important when we want to remove
   …… ions from …… water.

2 Write word equations to show what is formed when:

   a) nitric acid reacts with potassium hydroxide solution,
   b) lead nitrate solution reacts with potassium bromide solution.

KEY POINTS

1 An indicator is needed when we produce a salt by reacting an alkali with an acid to make a soluble salt.
2 Insoluble salts can be made by reacting two solutions to produce a precipitate.
3 Precipitation is an important way of removing some substances from wastewater.

C2 7.4   It's all in the soil

The importance of rotation

No, nothing to do with rotating YOU! This is about not growing the same vegetables in the same place two years running. If you do this you are likely to find two problems.

First, pests and diseases which live on the particular vegetables will increase, and you will have real problems.

Second, growing the same crop in the same place year after year will lead to the soil becoming unbalanced, with the level of some nutrients becoming too low.

Getting the right amount of acid

When the soil in your garden is too acid or alkaline, nutrients present in the soil become locked-up or unavailable. Acidic soil has a 'pH' that is too low (less than 7) while alkaline soil has a 'pH' that is too high. In fact, a decrease of just one pH unit means that the soil is ten times more acidic!

Getting the pH right is the same as applying fertiliser since it 'unlocks' plant nutrients which are already present.

Testing your soil

You can find out the pH of your soil by testing it with a simple soil testing kit. This will tell you how acidic or alkaline your soil is.

Follow the instructions in the kit, which usually involves mixing a little soil with some water and testing it with some special paper. The colour that the paper turns will tell you if your soil has a 'pH' that is too low or too high.

What to add . . . ?

If your soil has the wrong 'pH' you'll need to do something about it. Unless you live on chalky soil it's very unusual for the pH of soil to be too high. This is because adding fertiliser usually makes soil acidic. So the most common thing that you'll need to do every so often to keep your soil with a 'neutral pH' is to add lime.

Lime is made by heating limestone to decompose it. Lime reacts with acid in the soil, making it neutral. You can buy lime from your local garden centre.

ACTIVITY

Although there is a lot of chemistry in gardening, it is not often that it is explained clearly (or correctly!). Your job is to write an article for a gardening newspaper or a leaflet for a local garden centre.

It should describe the chemistry behind getting the pH of the soil correct by testing it and then adding the necessary chemicals. You can even use simple chemical equations (especially word equations) if this helps you to explain things more clearly.

Blue flag beaches

The idea of a way of showing clearly that a beach is clean was put forward in 1987, when 244 beaches from 10 countries were awarded a flag to show that the beach met certain standards. As far as water quality goes, these standards include:

* the cleanliness of the water must comply with the EU Bathing Water Directive,
* no industrial effluent or sewage discharges may affect the beach area,
* there must be local emergency plans to cope with pollution accidents.

In 2005 there were nearly 2500 blue flag beaches worldwide.

ACTIVITY

The town council of a seaside resort wishes to apply for a 'blue flag' for their beach. However, they have been told that they must get rid of a large amount of heavy metal pollution in the water discharged through the town's sewage system. The heavy metals come from a large factory near the town, which is a very important local employer.

Your job is to act as a consultant to the town council to advise them of the best way to go about cleaning up the effluent in order to be able to apply for a blue flag.

Write a report to the council explaining what they should do. You will need to explain the chemistry to them in simple terms (they should be able to understand simple word equations), and you will need to suggest who will pay for the treatment – whether this should be the local people (through their local taxes), the factory producing the pollution, or even the visitors to the town (through higher prices for their accommodation and other holiday costs).

Cleaning up industrial effluent

A lot of wastewater from industry contains salts of heavy (transition) metals dissolved in it. Before this can be discharged these must be removed. The simplest way of removing the metal ions is to raise the pH of the solution. The hydroxide ions in the alkaline solution then react with the metal ions, producing metal hydroxides. The hydroxides of most heavy metals are very insoluble. So these form a precipitate which can be removed from the wastewater before it is discharged into a river or into the sea.

SUMMARY QUESTIONS

1 Match the halves of the sentences together:

a) A base that is soluble in water	A ...... a pH of exactly 7.
b) Pure water is neutral with	B ...... form OH⁻ ions when they dissolve in water.
c) Acids are substances that	C ...... is called an alkali.
d) Alkalis are substances that	D ...... is acidic.
e) Indicators are substances that	E ...... produce H⁺ ions when they dissolve in water.
f) A solution with a pH less than 7	F ...... change colour when we add them to acids and alkalis.

2 The table shows the ions in substances in three pairs of beakers. Copy the table and draw lines between the ions that react in each beaker.

Beaker 1	Na^+ OH^-	H^+ Cl^-
Beaker 2	Cu^{2+} O^{2-}	H^+ H^+ SO_4^{2-}
Beaker 3	Pb^{2+} NO_3^- NO_3^-	Cu^{2+} Cl^- Cl^-

3 A student carried out an investigation in which she dropped a piece of magnesium ribbon into some acid. She measured the total amount of gas that had been produced every 10 seconds and plotted this on a graph. At the end of the reaction some magnesium ribbon remained that had not reacted.

a) What gas does this reaction produce?

b) Sketch a graph of volume of gas (*y*-axis) against time (*x*-axis) that this student could have obtained.

c) Sketch another line on the graph to show the results that might be obtained if the student repeated this investigation, using the same acid but diluted so that its concentration was half of that used in the first investigation.

4 Write chemical equations to describe the following chemical reactions. (Each reaction forms a salt.)

a) Potassium hydroxide (an alkali) and sulfuric acid.

b) Zinc oxide (an insoluble base) and nitric acid.

c) Calcium metal and hydrochloric acid.

d) Barium nitrate and sodium sulfate (this reaction produces an insoluble salt – **hint:** all sodium salts are soluble).

EXAM-STYLE QUESTIONS

1 Magnesium hydroxide, $Mg(OH)_2$, is used in many antacids for relieving acid indigestion.

(a) Magnesium hydroxide is slightly soluble in water.

(i) Give the formulae of the ions produced when it dissolves. (2)

(ii) Give a value for the pH of the solution it forms. (1)

(b) Write a word equation for the reaction of magnesium hydroxide with hydrochloric acid. (2)

(c) Write a balanced symbol equation for the reaction. (2)

(d) Suggest why sodium hydroxide would not be suitable for use as a cure for indigestion. (2)

2 Copper(II) sulfate crystals can be made from an insoluble base and sulfuric acid.

(a) Name the insoluble base that can be used to make copper(II) sulfate. (1)

(b) Describe how to make a solution of copper(II) sulfate from 25 cm³ of dilute sulfuric acid so that all of the acid is used. (3)

(c) Describe how you could make crystals of copper(II) sulfate from the solution. (3)

3 Salts are formed when acids react with alkalis.

(a) Complete the word equation:

   acid + alkali → .... + ... (2)

(b) What type of reaction takes place when an acid reacts with an alkali? (1)

(c) (i) Name the acid and alkali used to make potassium nitrate. (2)

(ii) What would you use to show when the acid had completely reacted with the alkali? (1)

(iii) Write a balanced symbol equation for the reaction that takes place. (2)

4 The effluent from nickel plating works is treated with sodium carbonate to precipitate nickel ions from the solution. The precipitate is separated from the solution by settlement in a tank. Filtration is not usually used as the main method of removing the precipitate, but can be used to remove small amounts of solids from the effluent after settlement.

(a) Write a word equation for the reaction between nickel sulfate solution and sodium carbonate solution. (2)

(b) Name the precipitate that is formed. (1)

(c) How is most of the precipitate removed from the effluent? (1)

(d) Suggest one reason why filtration is not used to remove most of the precipitate. (1)

(e) Why is it necessary to remove metal ions like nickel from effluents? (1)

5 There are four main methods of making salts:

A Acid + metal
B Acid + insoluble base
C Acid + alkali
D Solution of salt A + solution of salt B

(a) A student wanted to make some sodium sulfate.

  (i) Which method would be the best one to use? (1)

  (ii) Explain why you chose this method. (3)

  (iii) Name the reagents you would use. (2)

  (iv) Write a word equation for the reaction. (1)

(b) Another student wanted to make some magnesium carbonate.

  (i) Which method would you use for this salt? (1)

  (ii) Explain why you chose this method. (2)

  (iii) Name the reagents you would use. (2)

  (iv) Write a word equation for the reaction. (1)

HOW SCIENCE WORKS QUESTIONS

Chemistry to help!

Modern living produces enormous quantities of wastewater. Most of this can be treated in sewage works by biological processes. Sometimes biology cannot solve all of the problems. Chemistry is needed.

Phosphates are one such problem. You might have seen patches of stinging nettles near to old farms. These are due to the high concentrations of phosphates produced in animal (and human!) waste. This is a real problem when you have farms producing beef. In parts of USA they rear cattle with very little land.

The waste would normally be put on the land. It would cost a lot of money to transport the waste back to the farms that produced the feed. The waste is therefore dumped. The problem is that the high concentration of the phosphates causes water pollution problems.

Removing the phosphates at the sewage works by adding iron(III) chloride is also expensive. The wastewater therefore is treated with struvite, a magnesium salt that precipitates the phosphates.

Use these notes and your experience to answer these questions.

a) What are the economic issues associated with the disposal of animal waste? (2)

b) What are the environmental issues associated with the disposal of animal waste? (2)

c) What are the ethical issues associated with the use of chemistry to solve the problems of pollution? (1)

d) Who should be making the decisions about using chemistry to solve these problems? Explain your answer. (1)

e) Struvite can become a problem by precipitating out and blocking water pipes.
What extra information would be useful to those using struvite? (1)

EXAMINATION-STYLE QUESTIONS

1   Match these substances with the descriptions (a) to (e):

    diamond, hydrogen chloride, magnesium, neon, sodium chloride

    (a)   A compound made of small molecules.

    (b)   A gas at room temperature made of single atoms.

    (c)   A giant lattice of atoms that are covalently bonded.

    (d)   An ionic solid with a high melting point.

    (e)   A giant lattice that conducts electricity when it is solid.   *(5 marks)*

See pages 130–7

2   (a)   Draw a dot and cross diagram to show the electron arrangement of a lithium atom, atomic number 3.   *(2 marks)*

See pages 116–19

    (b)   Draw a dot and cross diagram to show the electron arrangement of a fluorine atom, atomic number 9.   *(2 marks)*

    (c)   Draw dot and cross diagrams to show the ions in lithium fluoride.   *(3 marks)*

3   Complete the table that shows information about some atoms.

See pages 114–21, 142

Symbol	Atomic number	Mass number	Number of protons	Number of neutrons	Electron arrangement of atom	Formula of ion	Electron arrangement of ion
Al	13	27	(a)	14	(b)	Al^{3+}	$[2,8]^{3+}$
O	8	16	8	(c)	2,6	O^{2-}	(d)
K	19	(e)	19	20	2,8,8,1	(f)	$[2,8,8]^{+}$
Cl	17	35	17	(g)	2,8,7	Cl^{-}	(h)

(8 marks)

4   A student added 20 g of marble chips to 50 cm³ of dilute hydrochloric acid in a conical flask. The flask was put onto a balance. The table shows the mass of gas that was given off. Some marble chips were left in the flask at the end of the reaction.

See pages 163–5

Mass of gas given off (g)	0	0.14	0.27	0.38	0.47	0.51	0.57	0.59	0.60
Time (minutes)	0	1.0	2.0	3.0	4.0	5.0	6.0	7.0	8.0

    (a)   Plot a graph of the results. Put time on the horizontal axis and mass lost on the vertical axis. Draw a smooth line through the points, omitting any result that is anomalous.   *(5 marks)*

    (b)   The rate of this reaction decreases with time. Explain how you can tell this from the graph.   *(1 mark)*

    The student decided to extend his work to see if temperature affected the rate at which the gas was produced.

    (c)   (i)   Suggest one control variable he should use.

GET IT RIGHT!

It is important to express yourself clearly in answers that require explanations. In Question 4(b), you should make it clear that the gradient or slope of the graph shows the rate of reaction at that time. Also, if you are asked how collisions affect the rate of reaction, it is not enough to say there are more collisions. It is the frequency of collisions (the number of collisions per second) that the rate depends upon.

(ii) Describe how he would control that variable. *(2 marks)*

(d) Suggest a suitable range of temperatures he could use. *(1 mark)*

(e) Suggest a suitable interval between temperatures. *(1 mark)*

(f) Use the first set of data to suggest a suitable length of time to leave the reaction. *(1 mark)*

5 Complete the table that shows information about the electrolysis of different substances. Carbon electrodes were used.

See pages 186–7

Substance	Positive ions present	Negative ions present	Product at negative electrode	Product at positive electrode
Molten magnesium chloride	Mg^{2+}	Cl^-	magnesium	(a)
Aqueous solution of potassium chloride	K^+ H^+	(b)	hydrogen	chlorine
Dilute sulfuric acid	(c)	SO_4^{2-} OH^-	hydrogen	oxygen
Aqueous solution of copper(II) sulfate	Cu^{2+} H^+	SO_4^{2-} OH^-	(d)	(e)

(5 marks)

6 Ammonium sulfate $(NH_4)_2SO_4$, is an important fertiliser. It is made by reacting ammonia solution with sulfuric acid. The reaction can be represented by the equation:

See pages 196, 200

$$H_2SO_4(aq) + 2NH_4OH(aq) \rightarrow (NH_4)_2SO_4(aq) + 2H_2O(l)$$

(a) How can you tell from the equation that ammonium sulfate is soluble?
(1 mark)

(b) (i) Which ions make the sulfuric acid solution acidic? *(1 mark)*

(ii) Which ions make the ammonia solution alkaline? *(1 mark)*

(iii) What name is used to describe the reaction between these ions? *(1 mark)*

(c) A student made 15.4 g of ammonium sulfate from 0.2 moles of sulfuric acid.

See pages 148–50

(i) What is the mass of one mole of ammonium sulfate? *(2 marks)*

(ii) What mass of ammonium sulfate can be made from 0.2 moles of sulfuric acid, according to the equation? *(1 mark)*

(iii) What was the percentage yield of ammonium sulfate obtained by the student? *(2 marks)*

[Higher]

GET IT RIGHT!

In calculations, always give your answer to an appropriate number of significant figures or decimal places (usually 2 or 3 significant figures or one or two decimal places).

C3 | Further chemistry

Filter beds like these are used to purify water

What you already know

Here is a quick reminder of previous work that you will find useful in this unit:

- The elements are shown in the periodic table and consist of atoms, which we can represent using chemical symbols.

- The properties of elements, compounds and mixtures can be used to classify them.

- Elements combine through chemical reactions to form compounds.

- We can represent compounds by formulae and we can summarise reactions by word equations.

- Displacement reactions take place between metals and solutions of salts of other metals.

- Metals and bases react with acids.

- Some everyday applications of neutralisation such as indigestion tablets.

- We can identify patterns in chemical reactions.

Lime is used to raise the pH of acidic soil

RECAP QUESTIONS

1 What elements do the following symbols represent?

a) C b) H c) N d) O e) Na f) Fe

2 What do we call the columns of elements in the periodic table?

3 Complete the word equation: sodium + chlorine →

4 What is an acid?

5 What is formed when alkalis react with acids?

6 What gas is produced when magnesium reacts with dilute hydrochloric acid?

7 A small volume of sodium chloride solution is left in a beaker on the window sill over the summer holidays. Explain what you will see when you get back after the holidays.

Making connections

Acids and bases around the home

Stinging nettles are covered with tiny hairs. When you touch a nettle the hairs stick into your skin. They inject an acid into you (called **oxalic acid**) – this is painful. (See the photo opposite.) It makes a tiny swelling as your body pumps liquid into the area to dilute the acid.

The traditional remedy for nettle stings is rubbing a dock leaf on the sting. But dock leaves are acidic too! It's likely that dock leaves help because rubbing the sting with them releases moist sap from the leaf which is cooling. But people hundreds of years ago had the right idea. They thought that goose dung was a good remedy! And what's so special about goose dung? It's alkaline!

That full-up, bloated feeling that you sometimes get when you've eaten a really big, rich meal can make you feel very uncomfortable. Sometimes the feeling doesn't go away, and you need something to 'settle your stomach'.

Indigestion tablets contain substances such as calcium carbonate. These react with the acid in your stomach to neutralise it and so put a stop to indigestion. Some of these substances produce carbon dioxide when they neutralise the acid – and this has to go somewhere! The quickest way out of the stomach is back up into your mouth . . . so you'd better be prepared to say 'pardon me!'

Baking powder is used to make cakes rise. It contains sodium hydrogencarbonate and tartaric acid. They are mixed together as dry powders so they do not react.

In the moist cake mixture the two chemicals react to produce carbon dioxide gas. Cooking the cake speeds up this reaction, and lots of gas bubbles form in the soft cake mixture. As the cake cooks the mixture hardens with the bubbles of gas trapped inside. This makes the cake light and tasty!

ACTIVITY

Chemistry is found all round the home – there are three examples here. But how many people really understand what's happening in these situations?

Design a page to go at the start of a kitchen book that will explain some of the chemistry of cooking, eating and gardening to people who will be using the book.

Chapters in this unit

Development of the periodic table   More about acids and bases   Water   Energy calculations   Analysis

C3 1.1   The early periodic table

LEARNING OBJECTIVES

1 Why do we need a periodic table?
2 What did the first periodic tables look like?

Imagine trying to understand chemistry:

● without knowing much about atoms,
● with each chemical compound having lots of different names, and
● without knowing a complete list of the elements.

Not an easy job! But that's the task that faced scientists at the start of the 1800s.

During the 19th century, chemists were finding new elements almost every year. At the same time they were trying very hard to find patterns in the behaviour of the elements. This would allow them to organise the elements and understand more about chemistry.

One of the first suggestions came from John Dalton, a teacher who lived most of his adult life in Manchester. Dalton arranged the elements in order of their mass, measured in various chemical reactions. In 1808 he published a table of elements in his book *A New System of Chemical Philosophy.* Look at his list in Figure 2.

a) How did Dalton put the elements in order?

In 1863, John Newlands built on Dalton's ideas with his *law of octaves* (an octave in music is eight notes). Newlands based this on the observation that the properties of every eighth element seemed similar.

He produced a table of his octaves, but he was so determined to make it work that he made several vital mistakes. He assumed that all the elements had been found – in spite of the fact that new ones were still turning up regularly. So he filled in his octaves regardless of the fact that some of his elements were not similar at all. He even put two elements in the same place at some points, to make everything fit in. So other scientists ridiculed his ideas, and they were not accepted.

b) What were the problems with Newlands' octaves?

Figure 1 Chemists' knowledge of the chemical elements in the early part of the 19th century was a bit like a partly-completed crossword puzzle – some things were clear, they had a vague idea about some elements, and they knew nothing about many more

Figure 2 Dalton and his table of elements

One year earlier a French chemist called Alexandre-Emile Beguyer de Chancourtois had already come up with a better attempt at sorting out the elements. He showed similarities between every eighth element and produced a clear diagram to demonstrate this. Unfortunately when his work was published the diagram was missed out!

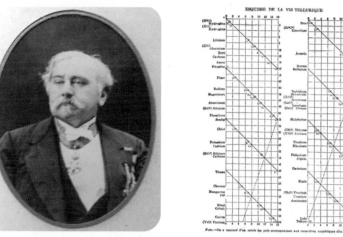

Figure 4 Without this diagram de Chancourtois' ideas were very difficult to understand, and so his work was largely ignored

Figure 3 Newlands and his table of octaves. Looking at Newlands' octaves, a fellow chemist commented that putting the elements in alphabetical order would probably produce just as many groups of elements with similar properties!

Then in 1869 the Russian scientist Dmitri Mendeleev cracked the problem. At this time 50 elements had been identified. Mendeleev arranged all of these in a table. He placed them in the order of their atomic masses. Then he arranged them so that a periodic pattern in their physical and chemical properties could be seen.

His stroke of genius was to leave gaps for elements which had not yet been discovered. But he predicted what their properties should be from his table. A few years later, new elements were discovered with properties which matched almost exactly Mendeleev's predictions.

Dmitri Mendeleev is remembered as the father of the modern periodic table.

Figure 5 Some of Mendeleev's original notes on the periodic table, together with a Russian stamp issued in his honour in 1969

SUMMARY QUESTIONS

1 Copy and complete using the words below:

discovered   gaps   mass   periodic   properties

The chemical elements can be arranged in a ...... table. Within the table, elements with similar ...... are placed together. Like other chemists, Mendeleev listed elements in order of ..., but he realised that he needed to leave ...... for elements that had not yet been ....

2 Mendeleev has an element named after him. What is it?

3 What was new about the way in which Mendeleev chose to order the elements in his table?

KEY POINTS

1 The periodic table of the elements developed as an attempt to classify the elements. It arranges them in a pattern according to their properties.
2 Early versions of the periodic table failed to take account of the fact that not all of the elements were known at that time.
3 Mendeleev's table took account of unknown elements, and so provided the basis for the modern periodic table.

C3 1.2

The modern periodic table

LEARNING OBJECTIVES

1 How do the elements fit into groups in the periodic table?
2 What are the patterns of behaviour within the groups?

DID YOU KNOW?

The alkali metal caesium reacts explosively with cold water, and reacts with ice at temperatures above −116°C.

Reactivity INCREASES going down the group – as the atoms get bigger the single electron in the outer energy level is attracted less strongly to the positive nucleus. This is because it is further away from the nucleus, and the inner shells of electrons screen it from the positive charge on the nucleus. The outer electron is easier to lose, so elements lower down the group are MORE reactive.

Group 1

Group 7

Reactivity DECREASES going down the group – as the atoms get bigger an electron added to the outer energy level is attracted less strongly to the positive nucleus. This is because it is further away from the nucleus, and the inner shells of electrons screen it from the positive charge on the nucleus. An extra electron is less easily attracted into the outer shell, so elements lower down the group are LESS reactive.

Using the relative atomic mass of elements was the only option available to Mendeleev at that time. His periodic table produced patterns which were recognised and accepted, but it had its limitations.

If we put the elements in order of the mass of their atoms, most elements end up in a group which behave in a similar way. But not all elements do this.

For example, argon atoms have a greater atomic mass than potassium atoms. This would mean that argon (a noble gas) would get grouped with extremely reactive metals such as sodium and lithium. And potassium (an extremely reactive metal) would be put with the noble gases. So argon would go in front of potassium in the periodic table, even though its atoms are heavier.

a) Find another pair of elements that get placed in the wrong order if atomic masses are used. (Refer to the periodic table on the next page.)

Once scientists began to find out more about atoms at the start of the 20th century, they could solve the problem described above. The structure of the atom was the key to developing the modern periodic table.

We now arrange the elements in order of their atomic (proton) number. This puts them all in exactly the right place in the periodic table. Their patterns of physical and chemical properties show this.

The periodic table is now a reliable model. It arranges the elements in groups with similar properties. It also provides us with an important summary of the structure of the atoms of all the elements.

b) How are elements arranged in the modern periodic table?

Going up and down the groups

Elements in the same group of the periodic table have similar properties. That's because their atoms have the same number of electrons in the highest occupied (outer) energy level.

Within a group, the properties of the elements are affected by the number of lower energy levels underneath the outer level. As we go down a group the number of occupied energy levels increases, and the atoms get bigger. This means that the outer electrons are further from the positive nucleus. There are also more energy levels between them and the nucleus.

This has two effects:

● the larger atoms *lose* electrons *more* easily,
● the larger atoms *gain* electrons *less* easily.

In both cases this happens because the outer electrons are further away from the nucleus. Not only that, the inner energy levels 'screen' or 'shield' the outer electrons from the positive charge in the nucleus. We can see this effect with the alkali metals and the halogens. (See the boxes on the left.)

Group numbers

1	2												3	4	5	6	7	0
						Atomic mass $\rightarrow$ 1 **H** 1 $\leftarrow$ Atomic number												4 **He** 2
7 **Li** 3	9 **Be** 4												11 **B** 5	12 **C** 6	14 **N** 7	16 **O** 8	19 **F** 9	20 **Ne** 10
23 **Na** 11	24 **Mg** 12												27 **Al** 13	28 **Si** 14	31 **P** 15	32 **S** 16	35.5 **Cl** 17	40 **Ar** 18
39 **K** 19	40 **Ca** 20	45 **Sc** 21	48 **Ti** 22	51 **V** 23	52 **Cr** 24	55 **Mn** 25	56 **Fe** 26	59 **Co** 27	59 **Ni** 28	63.5 **Cu** 29	65 **Zn** 30	70 **Ga** 31	73 **Ge** 32	75 **As** 33	79 **Se** 34	80 **Br** 35	84 **Kr** 36
85 **Rb** 37	88 **Sr** 38	89 **Y** 39	91 **Zr** 40	93 **Nb** 41	96 **Mo** 42	98 **Tc** 43	101 **Ru** 44	103 **Rh** 45	106 **Pd** 46	108 **Ag** 47	112 **Cd** 48	115 **In** 49	119 **Sn** 50	122 **Sb** 51	128 **Te** 52	127 **I** 53	131 **Xe** 54
133 **Cs** 55	137 **Ba** 56	139 **La** 57	178 **Hf** 72	181 **Ta** 73	184 **W** 74	186 **Re** 75	190 **Os** 76	192 **Ir** 77	195 **Pt** 78	197 **Au** 79	201 **Hg** 80	204 **Tl** 81	207 **Pb** 82	209 **Bi** 83	210 **Po** 84	210 **At** 85	222 **Rn** 86
223 **Fr** 87	226 **Ra** 88	227 **Ac** 89															

Elements 58–71 and 90–103 (all metals) have been omitted

Key

Reactive metals These metals react vigorously with other elements like oxygen or chlorine, and with water. They are all soft – some of them can even be cut with a knife, like cheese!

Transition metals This group contains the elements that most people probably think of when the word 'metal' is mentioned, like iron, copper, silver and gold. These metals are not usually very reactive – some, like silver and gold, are so unreactive that they are sometimes called 'noble metals'.

Non-metals These elements have low melting and boiling points, and many are liquids or gases at room temperature and pressure.

Noble gases These (non-metal) elements are very unreactive, and it is very difficult to get them to combine with other elements.

Figure 1 The modern periodic table. The upper number on the left of a symbol is the element's atomic mass (the number of protons and neutrons in its nucleus). The lower number is its atomic number (or proton number) – the number of protons in the nucleus.

Remember that the atoms of alkali metals tend to *lose* electrons when they form chemical bonds. On the other hand, the atoms of the halogens tend to *gain* electrons.

GET IT RIGHT!

Metals react by losing electrons. Non-metals react with metals by gaining electrons.

SUMMARY QUESTIONS

1 Copy and complete using the words below:

atomic   electrons   energy level   group   properties

The periodic table shows elements arranged in order of their ...... number. Elements in a ...... all have similar ...... because they have the same number of ...... in their outer ....

2 Where do we find the most reactive elements in the periodic table?

3 How do the number of metallic elements compare with the number of non-metal elements?

4 a) Explain why sodium is more reactive than lithium.
b) Explain why fluorine is more reactive than chlorine. [Higher]

KEY POINTS

1 The group that an element is in is determined by its atomic (proton) number.
2 The number of electrons in the highest energy level of an atom determines its chemical properties.
3 We can explain trends in reactivity as we go down a group in terms of the number of energy levels in the atoms. [Higher]

213

C3 1.3   Group 1 – the alkali metals

LEARNING OBJECTIVES

1 How do the Group 1 elements behave?
2 How do the properties of the Group 1 elements change with their position in the group?

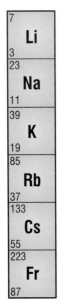

Figure 1 The alkali metals (Group 1)

We call the first group (Group 1) of the periodic table the **alkali metals**. This group consists of the metals lithium, sodium, potassium, rubidium, caesium and francium. The first three are the only ones you will usually have to deal with, as the others are extremely reactive. Francium is radioactive as well!

Properties of the alkali metals

All of the alkali metals are very reactive. They have to be stored in oil to stop them reacting with oxygen in the air. Their reactivity increases as we move down the group. So lithium is the least reactive alkali metal and francium the most reactive.

All of the alkali metals have a very low density for metals. In fact lithium, sodium and potassium will all float on water. The alkali metals are also all very soft, so we can cut them with a knife. They have the silvery, shiny look of typical metals when we first cut them. However, they quickly go dull as they react with oxygen in the air to form a layer of oxide.

The Group 1 metals also melt and boil at relatively low temperatures (for metals). As we go down the group, the melting points and boiling points get lower and lower.

The properties of this rather unusual group of metals are the result of their electronic structure. The alkali metals all have one electron in their highest energy level, which gives them similar properties. It also makes them very reactive, as they only need to lose one electron to obtain a stable electronic structure.

They react with non-metals, losing their single outer electron and forming a metal ion carrying a 1+ charge, e.g. Na^+, K^+. They always form ionic compounds.

a) What is the formula of a lithium ion?

Figure 2 The alkali metals have to be stored under oil

Figure 3 Lithium and potassium reacting with water (the lithium is on the right of the trough)

Reaction with water

When we add lithium, sodium or potassium to water the metal floats on the water, moving around and fizzing. The fizzing happens because the metal reacts with the water to form hydrogen gas. Potassium reacts so vigorously with the water that the hydrogen produced in the reaction catches fire. It burns with a lilac flame.

The reaction between an alkali metal and water also produces a metal hydroxide. The hydroxides of the alkali metals are all soluble in water, producing a colourless solution with a high pH. (Universal indicator turns purple.)

This is how the alkali metals got their name – they all form hydroxides which dissolve in water to give strongly alkaline solutions:

sodium + water → sodium hydroxide + hydrogen
2Na (s) + 2H₂O (l) → 2NaOH (aq) + H₂ (g)

$$2Na\ (s) + 2H_2O\ (l) \rightarrow 2NaOH\ (aq) + H_2\ (g)$$

potassium + water → potassium hydroxide + hydrogen
$$2K\ (s) + 2H_2O\ (l) \rightarrow 2KOH\ (aq) + H_2\ (g)$$

b) Write the word and symbol equations for the reaction of lithium with water.

Other reactions

The alkali metals also react vigorously with other non-metals such as chlorine. They produce metal chlorides which are white solids. These all dissolve readily in water to form colourless solutions.

These reactions get more and more vigorous as we go down the group. That's because it becomes easier to lose the single electron in the outer shell to form ions with a 1+ charge. (See page 212.)

sodium + chlorine → sodium chloride
$$2Na\ (s) + Cl_2\ (g) \rightarrow 2NaCl\ (s)$$

They react in a similar way with fluorine, bromine and iodine. All of the compounds of the alkali metals are ionic, so they form crystals. All of their compounds also dissolve easily in water.

DEMONSTRATION

Reactions of alkali metals with water

The reaction of the alkali metals with water can be explored by dropping small pieces of the metal into water. This must be done with great care, using a large volume and surface area of water. That's because the reaction is so vigorous, releasing a large amount of energy and hydrogen gas too.

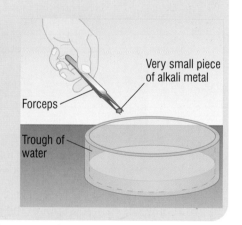

Forceps

Very small piece of alkali metal

Trough of water

SUMMARY QUESTIONS

1 Copy and complete using the words below:

alkali metals   alkaline   bottom   ionic   less   reactive   top

The Group 1 elements are also known as the .... They are very ......, producing ...... solutions when reacted with water. Lithium, at the ...... of the group, is ...... reactive than caesium, near the ...... of the group. These elements always form ...... compounds.

2 Francium is the element right at the bottom of Group 1.
What do you think would happen if it was dropped into water?

3 Write a balanced equation for the reaction of caesium with iodine. [Higher]

KEY POINTS

1 The elements in Group 1 of the periodic table are called the alkali metals.
2 The metals all react with water to produce hydrogen and an alkaline solution containing the metal hydroxide.
3 The reactivity of the alkali metals increases as we go down the group.

C3 1.4   Group 7 – the halogens

LEARNING OBJECTIVES

1 How do the Group 7 elements behave?
2 How do the properties of the Group 7 elements change with their position in the group?

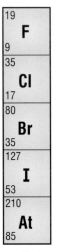

Figure 1 The Group 7 elements

Properties of the halogens

The halogens are a group of poisonous non-metals which all have coloured vapours. They are fairly typical non-metals:

- they have low melting points and boiling points,
- they are also poor conductors of heat and electricity.

The halogens all look different. At room temperature fluorine is a very reactive, poisonous, pale yellow gas, while chlorine is a reactive, poisonous dense green gas.

It is important to be able to detect chlorine if it is given off. It has a very distinctive smell – you'll recognise it from some swimming pools. But it is much safer to hold a piece of damp litmus paper in an unknown gas. If the damp litmus paper is bleached, the gas is chlorine.

Bromine is a dense, poisonous dark orange-brown liquid which vaporises easily – it is volatile. Iodine is a poisonous dark grey crystalline solid which produces violet-coloured vapour when we heat it.

As elements, the halogens all exist as molecules made up of pairs of atoms, joined together by covalent bonds.

	F — F (F_2)	Cl — Cl (Cl_2)	Br — Br (Br_2)	I — I (I_2)
Melting Point (°C)	−220	−101	−7	114
Boiling Point (°C)	−188	−35	59	184

Figure 2 The halogens all form molecules made up of pairs of atoms, joined by a covalent bond – we call this type of molecule a **diatomic** molecule

a) What patterns can you spot in the physical properties of the halogens as you go down Group 7?

Reactions of the halogens

The way the halogens react with other elements and compounds is a direct result of their electronic structure. They all have a highest energy level containing seven electrons. So they need just one more electron to achieve a stable arrangement.

This means that the halogens take part in both ionic and covalent bonding. It also explains why the halogens get less reactive as we go down the group. That's because the outer electrons get further away and are more and more shielded from the nucleus. (See page 212.)

b) Write a word equation for the reaction between hydrogen and fluorine.

FOUL FACTS

Many early chemists were badly hurt or even killed as they tried to make pure fluorine, which they called 'the gas of Lucifer'. It was finally produced by the French chemist Henri Moissan, who died aged just 55 – his life was almost certainly shortened by his work with fluorine.

	How the halogens react with hydrogen
$F_2 (g) + H_2 (g) \rightarrow 2HF (g)$	Explosive even at $-200°C$ and in the dark
$Cl_2 (g) + H_2 (g) \rightarrow 2HCl (g)$	Explosive in sunlight/slow in the dark
$Br_2 (g) + H_2 (g) \rightarrow 2HBr (g)$	$300°C$ + platinum catalyst
$I_2 (g) + H_2 (g) \rightleftharpoons 2HI (g)$	$300°C$ + platinum catalyst (very slow, reversible)

The halogens all react with metals. They gain a single electron to give them a stable arrangement of electrons, forming ions with a $1-$ charge, e.g. F^-, Cl^-, Br^-.

c) Write down the formula of an iodide ion.

In these reactions ionic salts, which we call *metal halides*, are formed. Some examples of these are sodium chloride ($NaCl$), iron(III) bromide ($FeBr_3$) and magnesium iodide (MgI_2). (See Figure 3.)

When we react the halogens with other non-metals, both sets of atoms share electrons to gain a stable electronic structure. Therefore their compounds with non-metals contain covalent bonds. (See Figure 4.)

Examples of these compounds are hydrogen chloride (HCl) and tetrachloromethane (CCl_4).

Displacement reactions between halogens

We can use a more reactive halogen to displace a less reactive halogen from solutions of its salts.

Bromine displaces iodine from solution because it is more reactive than iodine, while chlorine will displace both iodine and bromine.

For example, chlorine will displace bromine if we bubble the gas through a solution of potassium bromide:

$$Cl_2 + 2KBr \rightarrow 2KCl + Br_2$$

d) What would happen if bromine was added to a solution of potassium iodide?

Obviously fluorine, the most reactive of the halogens, would displace all of the others. However, it reacts so strongly with water that we cannot carry out any displacement reactions in aqueous solutions.

SUMMARY QUESTIONS

1 Copy and complete using the words below:

covalent   halogens   ionic   less   molecules   most   top   two

Group 7 elements are also called the .... Fluorine, at the ...... of the group, is the ...... reactive, while iodine is much ...... reactive. All of these elements exist as ...... made up of ...... atoms. They react with other non-metals to form ...... compounds and with metals to form ...... compounds.

2 The halogens react with the alkali metals to form ionic compounds. Chose any ONE alkali metal and any TWO halogens and write balanced equations for the possible reactions. [Higher]

GET IT RIGHT!

In Group 7, reactivity decreases as you go down the group but in Group 1 it increases going down the group.

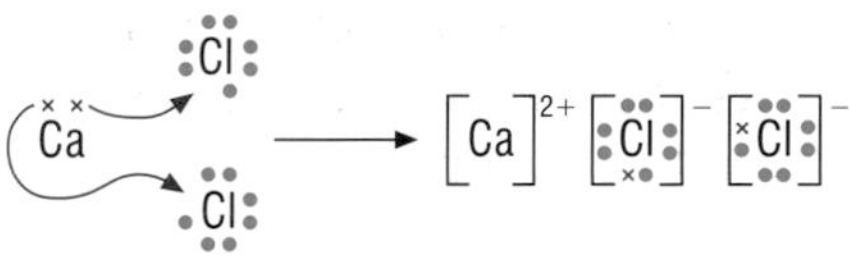

Figure 3 When we react a halogen and a metal, the metal donates electrons to the halogen and ionic bonds are formed between the ions produced

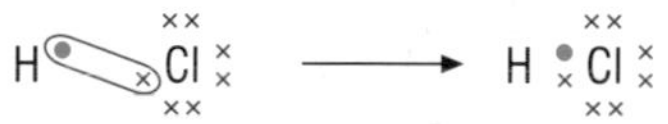

Figure 4 When we react a halogen with a non-metal they share electrons to form covalent bonds within the resulting molecules

PRACTICAL

Displacement reactions

Add bromine water to potassium iodide solution in a test tube. Then try some other combinations of solutions of halogens and potassium halides.

● Explain your observations.

KEY POINTS

1 The halogens exist as diatomic molecules.
2 The halogens all form ions with a single negative charge.
3 The halogens form covalent compounds by sharing electrons with other non-metals.
4 The reactivity of the halogens decreases going down the group.

C3 1.5

The transition elements

LEARNING OBJECTIVES

1 What are the properties of the transition elements?
2 How do the transition elements react?

In the centre of the periodic table there is a large block of metallic elements. We call these the **transition metals** or **transition elements**.

45 Sc 21	48 Ti 22	51 V 23	52 Cr 24	55 Mn 25	56 Fe 26	59 Co 27	59 Ni 28	63 Cu 29	64 Zn 30
89 Y 39	91 Zr 40	93 Nb 41	96 Mo 42	99 Tc 43	101 Ru 44	103 Rh 45	106 Pd 46	108 Ag 47	112 Cd 48
	178 Hf 72	181 Ta 73	184 W 74	186 Re 75	190 Os 76	192 Ir 77	195 Pt 78	197 Au 79	201 Hg 80

Figure 1 The transition metals. The more common metals are shown in **bold type**.

Physical properties

Many of the transition elements have similar properties, but they are also different from the other elements in the periodic table.

The transition metals have a typical metallic structure which explains most of their properties. The metal's atoms exist in a giant structure held together by metallic bonds, and the outer electrons of each atom can move about freely within the metal. (See page 124.)

Like all metals, the transition metals are very good conductors of electricity and heat because delocalised electrons carry the electrical current or the heat energy through the metal. The transition metals are also hard, tough and strong, yet we can bend or hammer them into useful shapes.

a) What word do we use to describe metals that can be hammered into shapes without shattering?

With the exception of mercury, which is a liquid at room temperature, the transition metals have very high melting points. This is very clear if we compare them to the alkali metals of Group 1. (See Figure 2.)

Many of the properties of the transition elements are due to the arrangement of electrons in their atoms. In these elements a lower energy level (or inner shell) is filled up between Groups 2 and 3. This partly-filled lower energy level explains why transition metals form brightly coloured compounds. It also results in their use as catalysts.

HIGHER

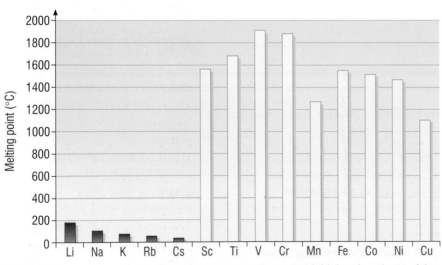

Figure 2 The melting points of the transition elements are much higher than those of the Group 1 elements

Chemical properties and alloys

The transition elements are much less reactive than the metals in Group 1. This means they do not react as easily with oxygen or water as the alkali metals. In other words, they corrode very slowly. Together with their physical properties, this makes the transition elements very useful as structural materials.

They are particularly useful when they are mixed together with each other or with other elements to make **alloys**. Iron mixed with carbon in steels is certainly the best known of these. (See page 42.)

Other very useful alloys of transition elements are:

- *brass*, a combination of copper and zinc, and
- *cupro-nickel*, a very hard alloy of copper and nickel which is used to make the coins we use in our currency.

Transition metal compounds

Many of the transition metals form coloured compounds. These include some very common compounds that we use in the laboratory. For example, potassium dichromate(VI) is orange – the orange colour is due to the chromium ion in the compound. Similarly copper(II) sulfate is blue (from the copper ion) and potassium manganate(VII) is purple (from the manganese ion).

b) Write down the full name of this transition metal compound, $NiCl_2$.

The colours which are produced by the transition elements are important in the world around us. For example, the colours of many minerals, rocks and gem stones are the result of transition element ions.

A reddish-brown colour in a rock is often the result of iron ions. The blue colour of sapphires and the green of emeralds are both due to transition element ions in the structure of the crystal.

We use the coloured ions of the transition elements in many different ways. Lots of pottery glazes contain transition metal ions to give bright colours. As copper weathers it produces a green film of basic copper carbonate. This green patina is very attractive and is generally called 'verdigris'. It is one reason why copper is used for many statues.

SUMMARY QUESTIONS

1 Copy and complete using the words below:

coloured	conductors	densities	less	melting
	three		two	

The transition metals are a block of elements that lie between Groups ...... and ...... in the periodic table. These elements nearly all have high ......, high ...... points and are good .... They are ...... reactive than the alkali metals, and often form ...... compounds.

2 Iron (Fe) can form ions that carry a 2+ or a 3+ charge. Write down a) the formula and b) the full names of the compounds iron can form with chlorine.

3 Vanadium (symbol V) reacts with oxygen gas to form the compound vanadium(V) oxide. Write a balanced equation for the reaction. [Higher]

DID YOU KNOW?

Coins are made from transition elements that scientists have found to be 'self-sterilising' – bacteria on the surface of the coins are poisoned by the metals, making it unlikely that diseases will be transmitted in this way.

When we write down the formula of a compound containing a transition element we usually include a Roman number as well as the name of the compound, e.g. **potassium manganate(VII)**, **copper(II) sulfate**.

This is because transition elements have more than one ion. For example, iron may exist as Fe^{2+} or Fe^{3+}, copper as Cu^+ and Cu^{2+}, and chromium as Cr^{2+} and Cr^{3+}. These ions are often very brightly coloured. For example, iron(II) ions (Fe^{2+}) give compounds a green colour, while iron(III) (Fe^{3+}) ions produce the reddish-brown colour we associate with rust!

GET IT RIGHT!

The charge on the ion is given in the name of the transition metal compound, e.g. copper(II) sulfate.

KEY POINTS

1 Nearly all the transition elements have high melting points and high densities.
2 The transition metals are strong and hard, and are good conductors of electricity and heat.
3 The transition metals do **not** react vigorously with oxygen or water.

C3 1.6   Finding and creating new elements

CHEMISTRY IN CHAOS

Karlsruhe, Germany, 3 September 1860

There is a meeting of the International Chemical Congress today at the suggestion of internationally-renowned chemist Friedrich August Kekulé. It aims to decide some clear and simple ways in which the formulae of compounds can be represented. At the moment different people use different formulae for the same chemical compound, which makes it very difficult for scientists to communicate their ideas. The conference will be addressed by Stanislao Cannizzaro who will talk about the importance of atomic masses.

The Law Of Octaves

London, Chemical News, 20 August 1864

Mr John Newlands writes:

Sir, – In addition to the fact stated in my late communication, may I be permitted to observe that if the elements are arranged in the order of their equivalents, calling hydrogen 1, lithium 2, glucinum 3, boron 4, and so on (a separate number being attached to each element having a distinct equivalent of its own, and where two elements happen to have the same equivalent, both being designated by the same number), it will be observed that elements having consecutive numbers frequently either belong to the same group or occupy similar positions in different groups …… the difference between the number of the lowest member of a group and that immediately above it is 7; in other words, the 8th element starting from a given one, is a kind of repetition of the first, like the eighth note of an octave of music.

A New Periodic Table

Moscow, 6 March 1869

A presentation was made by Professor Menshutken to the Russian Chemical Society today on behalf of his colleague, Dmitri Mendeleev. He described an arrangement of the elements in which the elements, if arranged according to their atomic weights, exhibit an apparent periodicity of properties. Professor Mendeleev also suggests that 'elements which are similar as regards their chemical properties have atomic weights which are either of nearly the same value (e.g. Pt, Ir, Os) or which increase regularly (e.g. K, Rb, Cs)' and that 'we must expect the discovery of many as yet unknown elements – for example, elements analogous to aluminium and silicon – whose atomic weight would be between 65 and 75.'

ACTIVITY

Design a poster to advertise a talk to be given by Dmitri Mendeleev about his new periodic table. In your poster describe these new ideas clearly, and say why they are important.

```
Nature Magazine
Back  Forward  Stop  Refresh  Home   Autofill  Print  Mail
Address:
```

MODERN ALCHEMISTS MAKE TWO NEW ELEMENTS

3 February, 2004

Tantalising evidence of two new chemical elements has been produced by a team of Russian and American scientists. Their observations indicate that we may be getting close to the fabled 'island of stability' in the periodic table, where heavy elements should be more stable than their neighbours. If confirmed, the discovery will bring the tally of known elements to 116.

Uranium, the heaviest element found in nature, has an atomic number of 92, meaning it has 92 protons in its nucleus. Atoms bigger than this are more likely to split apart spontaneously in radioactive decay. That's because the strong nuclear force that holds protons and neutrons together gets weaker as more particles jostle for space at the core of the atom. Also, protons have a positive charge and the more there are, the greater the strain on the nucleus due to the repulsion between them. Eventually the nucleus shatters, spraying out smaller, more stable atoms.

But physicists have predicted 'islands of stability' at atomic numbers 114, 120 and/or 126, where the protons and neutrons might be able to jostle themselves into a shape that minimises contact between the protons. That would allow the nucleus to hang together for much longer than its neighbours in the periodic table. Creating such elements may give scientists access to unusual and exciting chemistry.

```
Chemistry World
Back  Forward  Stop  Refresh  Home   Autofill  Print  Mail
Address:
```

Exploring the Outer Reaches

June 2005

- Of the 117 chemical elements that are currently known, or have been claimed to exist, 29 have been synthesised in laboratories.

- Of these, elements 113 and 115 have been synthesised but have not been verified. Element 118 has also been claimed but not yet confirmed.

- Synthesising elements higher than 118 will be difficult but hopes are pinned on the existence of 'islands of stability' for atomic nuclei which may make it possible to synthesise elements with atomic numbers as high as 130.

ACTIVITY

Synthesising new elements is expensive, and requires a huge investment in equipment and scientists. Imagine that you have been asked to present a radio programme about the search for new elements. Working in groups, write a script for the programme, including interviews with chemists working in this area. You may wish to make your programme present a balanced point of view – or different groups can argue for one side only. When you have written your script it can be recorded or performed for the rest of the class.

SUMMARY QUESTIONS

1 Link each scientist (A–C) with his idea (1–3).

A	Newlands	1	Arranged the elements in order of their masses.
B	Mendeleev	2	Produced the 'law of octaves'.
C	Dalton	3	Produced the periodic table on which our modern table is based.

2 What is the difference between a **group** and a **period** in the periodic table?

3 Write down word equations for the following reactions:

a) sodium and water

b) caesium and chlorine

c) fluorine and aluminium

d) chlorine and hydrogen

e) bromine and potassium iodide solution.

4 Write down balanced chemical equations for the reactions in question 3. [Higher]

5 Draw dot and cross diagrams (draw the outer energy shell electrons only) to show the bonding in the following substances:

a) hydrogen fluoride

b) chlorine

c) sodium iodide.

6 a) How are the properties of the transition elements different to the properties of the alkali metals?

   b) Why are the transition elements more useful than the alkali metals as structural materials (materials used to make things)?

7 Most chemists consider that zinc is not a transition element. Find out more about the properties of zinc and its compounds and suggest why zinc may not be called a transition element. [Higher]

EXAM-STYLE QUESTIONS

Use a periodic table (on page 213 of Student Book or on the AQA Data Sheet) to help you to answer these questions.

1 Choose elements from the list to match the descriptions (a) to (e).

> **aluminium** **astatine** **bromine** **caesium**
> **chromium** **fluorine** **lithium** **magnesium**

(a) A liquid at room temperature with a brown vapour. *Br*

(b) A metal that floats on water. *Mg*

(c) A transition element. *Cr*

(d) The most reactive element in Group 1. *Cs*

(e) The most reactive element in Group 7. *Ft* (5)

2 Some students observed sodium reacting with water. They wrote down these observations.

(a) The metal turned into a silver ball.

(b) The metal moved around on the surface of the water. *highly very reactive*

(c) A fizzing sound was heard. *H₂ produced*

(d) The metal gradually disappeared. *reacted w/ H₂O fully*

Write down one deduction you can make from each of these observations. (4)

3 Part of Mendeleev's periodic table is shown below.

	Group 1	Group 2	Group 3	Group 4	Group 5	Group 6	Group 7	Group 8
Period 1	H							
Period 2	Li	Be	B	C	N	O	F	
Period 3	Na	Mg	Al	Si	P	S	Cl	
Period 4	K Cu	Ca Zn	? ?	Ti ?	V As	Cr Se	Mn Br	Fe Co Ni
Period 5	Rb Ag	Sr Cd	Y In	Zr Sn	Nb Sb	Mo Te	? ?	Ru Rh Pd

(a) (i) Write the symbols of the three elements in Group 1 of Mendeleev's table that are not in Group 1 of the modern periodic table. (1)

   (ii) Where are these elements placed in the modern table? (2)

   (iii) Suggest one reason why Mendeleev put all of these elements in the same group. (1)

(b) Which group of elements in the modern periodic table is completely missing from Mendeleev's table? (1)

(c) (i) In what order did Mendeleev place the elements in his table? (1)

　　(ii) In what order are elements in the modern periodic table? (1)

(d) Explain why Mendeleev left the gaps shown by question marks (?) in his table. (2)

4 The table shows the numbers of protons and neutrons in ten different atoms.

Atom	A	E	G	J	L	M	Q	R	T	X
Number of protons	9	11	12	17	17	18	19	20	22	24
Number of neutrons	10	12	12	18	20	22	20	20	26	28

(a) Which are atoms of elements in Group 1?

(b) Which is an atom of a noble gas?

(c) Which are atoms of transition elements?

(d) Which two atoms are isotopes of the same element?

(e) Which atoms have the same mass number?

(f) Write the formula of the ion formed by atom E.

(g) Write the formula of the ion formed by atom L.

(h) Write the formula of the compound formed when A reacts with R. (8)

5 This question is about the halogens, Group 7 in the periodic table.

(a) Write down one similarity and one difference in the electronic structures of the elements going down Group 7 from fluorine to astatine. (2)

(b) How does the reactivity of the halogens change going down the group? (1)

(c) Explain this change in reactivity in terms of atomic structure. (3)

[Higher]

HOW SCIENCE WORKS QUESTIONS

Tony is starting an investigation into the halogens. He has been told about the order in which different halogens appear in the periodic table and from that has devised a method for seeing which is the most reactive. He has chosen you as his partner! It all sounds very complex. Tony says he is going to use four test tubes.

In tube 1 he is going to put a few crystals of potassium iodide into some chlorine water – a colour change would prove that chlorine is more reactive than iodine.

In tube 2 he is going to put a few crystals of potassium iodide into some bromine water – a colour change would prove that bromine is more reactive than iodine.

In tube 3 he is going to put a few crystals of potassium chloride into some bromine water – a colour change would prove that bromine is more reactive than chlorine.

In tube 4 he is going to put a few crystals of potassium bromide into some chlorine water – a colour change would prove that chlorine is more reactive than bromine.

Tony quickly carries out the first reaction and then decides that it would be a good idea if you produced a table for his results.

a) Make a table to show what is being mixed and leave room for the results. (3)

b) The results were that tubes 1 and 2 went dark brown. Tube 3 showed no change of colour and tube 4 went yellowy in colour. Now complete your table with these results. (1)

c) You told Tony that you thought 'it could have been the water that changed the first two dark brown'. What could he do to prove that it was not the water? (1)

d) Is the dependent variable categoric, continuous, discrete or ordered? (1)

e) Can you see a pattern from these results? If so, state the pattern. (1)

C3 2.1   Strong and weak acids/alkalis

LEARNING OBJECTIVES

1 How is water involved when acids and bases react?
2 What ions are responsible for acidity and alkalinity?
3 What are strong and weak acids?

We saw on page 196 that acids form H^+ ions when they dissolve in water. An H^+ ion is a hydrogen atom which has lost its electron – in other words, it is simply a proton.

A proton produced as we dissolve an acid in water becomes surrounded by water molecules which keep it in solution – we say that it is **hydrated**. Hydrated hydrogen ions are represented as H^+ *(aq)*.

An alkali is a base which dissolves in water. When a base dissolves in water it produces OH^- *ions* (hydroxide ions).

HIGHER

Because acids act as a source of protons, we often refer to them as **proton donors**.

a) What is another name for an acid?

The hydroxide ions from an alkali (soluble base) combine readily with protons (H^+ ions) to form water:

$$OH^- (aq) + H^+ (aq) \rightarrow H_2O (l)$$

Because they behave like this, we often call bases **proton acceptors**.

b) What is another name for a base?

Strong and weak acids

We can classify acids and alkalis as 'strong' or 'weak'. This depends on the extent to which they ionise in water.

A strong acid or alkali is one that is 100% ionised in water. Hydrochloric, sulfuric and nitric acids are all examples of strong acids. Sodium hydroxide and potassium hydroxide are examples of strong alkalis.

On the other hand, a weak acid or alkali is only partly ionised in water. Examples of weak acids are ethanoic, citric and carbonic acid. Weak alkalis include ammonia solution.

c) Give two examples of strong acids and two examples of strong alkalis.
d) Give two examples of weak acids and one example of a weak alkali.

How can we tell if an acid is weak or strong? There are two main ways. The pH scale is a measure of the concentration of hydrogen ions in a solution. So if we measure the pH of a weak acid and a strong acid which have the same concentration, the strong acid will have a lower pH. That's because it will be fully ionised.

GET IT RIGHT!

Remember that **strong** and **concentrated** are **not** the same thing – neither are **weak** and **dilute**.

pure hydrogen chloride
(Strong acid)

dissolve each acid in water to make
a 1 mol/dm³ solution →

completely ionised, so concentration of H^+ ions $= 1\,mol/dm^3$

pure citric acid
(Weak acid)

only partly ionised, so concentration of H^+ ions much less than $1\,mol/dm^3$

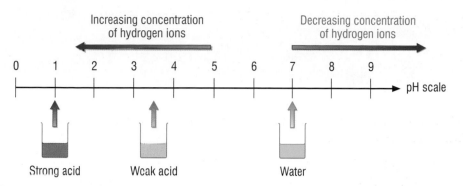

Figure 1 When we compare two acids which have the same concentration, the *stronger acid* will have the *lower pH*.

Another way that we can distinguish between weak and strong acids which have the same concentration is by observing the rate of reaction when we add a reactive metal to the acid.

PRACTICAL

Comparing a strong and a weak acid

If you add a piece of metal such as magnesium or zinc to a strong acid we get hydrogen gas given off. You can record the rate of reaction by observing the rate at which hydrogen gas is produced.

If we now carry out the reaction using a weak acid which has the same concentration as the strong acid, the reaction is much slower.

acid	Observations
Hydrochloric acid, 1 mol/dm³	Fizzed quite violently, magnesium ribbon had disappeared in 5 seconds.
Citric acid, 1 mol/dm³	Reacted very slowly, bubbles of gas produced, magnesium ribbon still visible after 90 seconds of reaction.

1 mol/dm³

Weak acid

1 mol/dm³

Strong acid

SUMMARY QUESTIONS

1 Copy and complete using the words below:

   alkali   hydrogen   hydroxide   protons   water

   When we dissolve an acid in water, …… ions or …… are produced. An …… is a base which dissolves in water. When it does this, it produces …… ions.
   H⁺(aq) and OH⁻(aq) ions react together to form …… .

2 Describe *one* way in which you could distinguish between a weak acid and a strong acid, both with the same concentration.

3 A student reacted some magnesium ribbon with a weak acid and measured the amount of gas produced. At the end of the reaction there was no magnesium ribbon left.
   a) What gas is produced? Describe the test for this gas.
   b) Sketch a graph showing the volume of gas produced against time.
   c) Sketch another line on this graph to show what she might observe if she repeated the investigation using a strong acid and the same mass of magnesium ribbon.

KEY POINTS

1 Acids in aqueous solutions produce H⁺ ions.
   Alkalis in aqueous solutions produce OH⁻ ions.
2 A strong acid or base is 100% ionised in water.
3 A weak acid or alkali is only partly ionised in water.
4 An acid is a proton donor.
   A base is a proton acceptor.
   [Higher]

C3 2.2   Titrations

LEARNING OBJECTIVES

1 How can we measure accurately the amount of alkali needed to react completely with an acid?
2 How do we know when the reaction is complete?

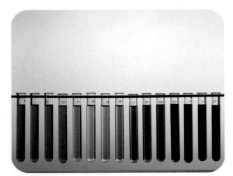

Figure 1 An indicator is a chemical that changes colour to show that a reaction has happened. Universal indicator is a mixture of indicators.

So far we have seen that adding an acid solution to an alkaline solution will produce a neutralisation reaction. The acid and the alkali react together and neutralise each other, forming a salt in the process.

Suppose we carry out a neutralisation reaction with a strong acid and a strong alkali. The solution that is produced will be neutral only if there are exactly the right quantities of acid and alkali in the solutions to start with.

If we start off with more acid than alkali, then although all the alkali will be neutralised the solution that is formed will be acidic. That's because some acid will be left over (in excess).

a) If there is more acid than alkali to start with in a neutralisation reaction, what will happen?

On the other hand, if we have more alkali than acid to begin with, then all the acid will be neutralised and the solution produced will be alkaline.

b) If there is more alkali than acid to start with in a neutralisation reaction, what will happen?

We can measure the precise volumes of acid and alkali solutions needed to react with each other by a technique which we call **titration**. The point at which the acid and alkali have reacted completely is called the **end point** of the reaction. We show this by using a chemical called an **indicator**.

c) What do we call the chemical used to show the end point of a neutralisation reaction?

Choosing an indicator

The pH of the solution left when an acid and alkali have completely reacted is not always 7. A neutral solution only forms when exactly the right amounts of a strong acid react with a strong alkali. If a strong acid reacts completely with a weak alkali, the solution formed at the end point is acidic. On the other hand, if a strong alkali reacts completely with a weak acid, the solution formed at the end point is alkaline.

Indicators change colour over different pH ranges. So you have to choose a suitable indicator when carrying out titrations with different combinations of acids and alkalis.

For example,

● strong acid + strong alkali – any indicator is suitable,

● weak acid + strong alkali – use phenolphthalein,
● strong acid + weak alkali – use methyl orange.

GET IT RIGHT!

We use a **pipette** to measure out a fixed volume of solution.

We use a **burette** to measure the volume of the solution added.

PRACTICAL

Carrying out a titration

To carry out a titration to find out how much acid is needed to completely react with a solution of alkali we use the following method:

1 Measure a known volume of the solution of alkali into a conical flask using a **pipette**. (Before you do this you should first wash the pipette with clean distilled water, and then with some of the solution.)

2 Now add an indicator solution to the solution in the flask.

3 Put a solution of the acid you are going to use into a **burette**. This is a long tube with a tap on one end. The tube has markings on it to enable you to measure accurately the volume of solution let out of the burette – often to the nearest 0.05 cm³. (Before you do this you should first wash the burette with clean distilled water, and then with some of the acid solution.)

4 Record the reading on the burette, then open the tap to release a small amount of acid into the flask. Swirl the flask to make sure that the two solutions are mixed.

5 Keep on repeating step 4 until the indicator in the flask changes colour, showing that the alkali in the flask has completely reacted with the acid added from the burette. Record the reading on the burette and calculate the volume of acid run into the flask. (On your first go at doing this you will probably run too much acid into the flask, so treat this as a rough estimate of how much acid is needed.)

6 Repeat the whole process at least three times. Discard any anomalous results. (See page 11.) Then calculate an average value to give the most reliable results possible.

7 Now you can use your results to calculate the concentration of the alkali in the solution. (See pages 228 and 229.)

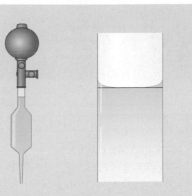

Figure 2 A pipette and pipette filler. Fill the pipette until the bottom of the meniscus coincides with the mark. Allow the liquid to run out of the pipette and touch the tip on the side of the flask to drain out the liquid – it is normal for a tiny amount of liquid to remain in the pipette.

Figure 3 A burette – use the bottom of the meniscus to read the scale. The reading here is 0.65 cm³.

SUMMARY QUESTIONS

1 Copy and complete using the words below:

   acid   end point   indicator   neutralisation

   Adding an acid to an alkali produces a ...... reaction. The point at which enough ...... has been added to completely react with the alkali is called the .... This can be shown using a chemical called an ....

2 Draw a flowchart to show how to carry out a titration between an acid of unknown concentration and an alkali of known concentration.

3 Indicator A changes from deep blue in acid solution to colourless in alkaline solution. Indicator B changes from pale green in acid solution to pale blue in alkaline solution. Which is the better indicator to use in an acid–base titration and why?

KEY POINTS

1 Titration is used to measure accurately how much alkali is needed to react completely with a known amount of acid (or vice versa).

2 The point at which an acid–base reaction is complete is called the end point of the reaction.

3 A suitable indicator should be chosen to show the end point of an acid–base reaction.

[Higher]

C3 2.3   Titration calculations

LEARNING OBJECTIVES

1 How can we calculate concentrations from reacting volumes?

2 How can we calculate the amount of acid or alkali needed in a neutralisation reaction?

We describe the concentration of a solute in a solution in terms of the number of moles of solute dissolved in one cubic decimetre of solution. We write these units as **moles per cubic decimetre** or **mol/dm³** for short. So if we know the amount of substance dissolved in a certain volume of a solution, we can work out the concentration of the solution.

As an example, imagine that we make a solution of sodium hydroxide in water by dissolving exactly 40 g of sodium hydroxide to make exactly 1 dm³ of solution. We know that the mass of 1 mole of sodium hydroxide (NaOH) is the sum of the atomic masses of sodium, oxygen and hydrogen, i.e. $23 + 16 + 1 = 40$ g. So we know that the solution contains exactly 1 mole of sodium hydroxide in exactly 1 dm³ of solution – and the concentration of sodium hydroxide in the solution is therefore 1 mol/dm³.

Worked example 1

But what if we use exactly 40 g of sodium hydroxide to make exactly 500 cm³ of solution instead of 1 dm³? (Remember that 1 dm³ = 1000 cm³.)

Solution

To find the concentration of the solution we must work out how much sodium hydroxide there would be in exactly 1000 cm³ (1 dm³) of solution if the proportions of sodium hydroxide and water stayed the same.

40 g of NaOH are dissolved in 500 cm³ water, so

$\dfrac{40}{500}$ g of NaOH would be dissolved in 1 cm³ of solution, and

$\dfrac{40}{500} \times 1000$ g = **80 g** of NaOH would be dissolved in 1000 cm³ of solution.

The mass of 1 mole of NaOH is 40 g,
so 80 g of NaOH is $80 \div 40$ moles = **2 moles**.

2 moles of NaOH are dissolved in 1 dm³ of solution. So the concentration of NaOH in the solution is **2 mol/dm³**.

Sometimes we know the concentration of a solution and need to work out the mass of solute in a certain volume.

Worked example 2

What mass of H_2SO_4 is there in 250 cm³ of 1 mol/dm³ sulfuric acid (H_2SO_4)?

Solution

In 1 dm³ of acid there would be 1 mole of H_2SO_4

The mass of 1 mole of H_2SO_4 is $(2 \times 1) + 32 + (4 \times 16)$ g = 98 g, so

in 1000 cm³ of solution there would be 98 g of H_2SO_4, and

in 1 cm³ of solution there are $\dfrac{98}{1000}$ g of H_2SO_4, so

in 250 cm³ of solution there are $\dfrac{98}{1000} \times 250$ g of H_2SO_4 = **24.5 g** of H_2SO_4

There is **24.5 g of H_2SO_4** in 250 cm³ of 1 mol/dm³ sulfuric acid.

Titration calculations

In a titration we always have one solution with a concentration which we know accurately. We put this in the burette. We put the other solution, which contains a known substance but with an unknown concentration, in a conical flask. We do this using a pipette to ensure that we know the volume of this solution in the flask. The result from the titration is used to calculate the number of moles of the substance in the solution in the conical flask.

Worked example 3

A student put 25.0 cm³ of sodium hydroxide solution with an unknown concentration into a conical flask using a pipette. The sodium hydroxide reacted with exactly 20.0 cm³ of 0.50 mol/dm³ hydrochloric acid added from a burette. What was the concentration of the sodium hydroxide solution?

Solution

The equation for this reaction is:

$$NaOH\ (aq) + HCl\ (aq) \rightarrow NaCl\ (aq) + H_2O\ (l)$$

This equation tells us that 1 mole of NaOH reacts with 1 mole of HCl.

The concentration of the HCl is 0.50 mol/dm³, so

0.50 moles of HCl are dissolved in 1000 cm³ of acid, and

$\dfrac{0.50}{1000}$ moles of HCl are dissolved in 1 cm³ of acid, therefore

$\dfrac{0.50}{1000} \times 20.0$ moles of HCl are dissolved in 20.0 cm³ of acid.

There are 0.010 moles of HCl dissolved in 20.0 cm³ of acid.

The equation for the reaction tells us that 0.010 moles of HCl will react with exactly 0.010 moles of NaOH. This means that there must have been 0.010 moles of NaOH in the 25.0 cm³ of solution in the conical flask. To calculate the concentration of NaOH in the solution in the flask we need to calculate the number of moles of NaOH in 1 dm³ of solution.

0.010 moles of NaOH are dissolved in 25.0 cm³ of solution, so

$\dfrac{0.010}{25}$ moles of NaOH are dissolved in 1 cm³ of solution, and there are

$\dfrac{0.010}{25} \times 1000 = 0.40$ moles of NaOH in 1000 cm³ solution.

The concentration of the sodium hydroxide solution is 0.40 mol/dm³.

SUMMARY QUESTION

1 In a titration, a 25.0 cm³ sample of nitric acid (HNO_3) reacted exactly with 20.0 cm³ of 0.40 mol/dm³ sodium hydroxide solution.

   a) Write down a balanced equation for this reaction.
   b) Calculate the number of moles of sodium hydroxide added.
   c) Write down the number of moles of HNO_3 in the acid.
   d) Calculate the concentration of the nitric acid.

200 cm³ of sodium hydroxide solution.
Volume of 1.0 mol dm⁻³ hydrochloric acid needed to neutralise it:

Expt 1: 10.1 cm³
Expt 2: 10.2 cm³
Expt 3: 10.3 cm³

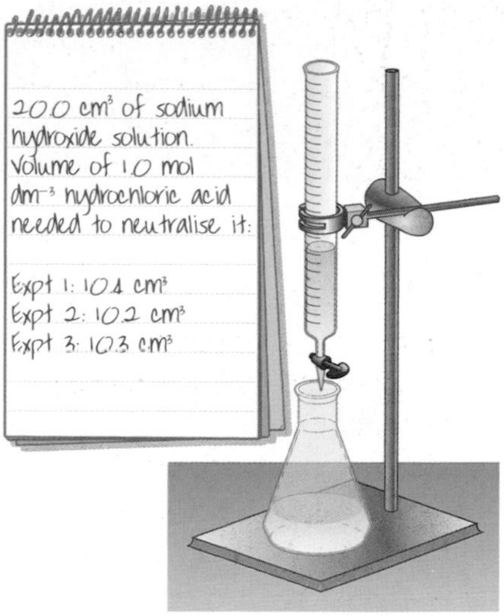

Figure 1 From results like these we can work out the concentration of the unknown solution

KEY POINTS

1 To calculate the concentration of a solution
   ● calculate the mass (in grams) of solute in 1 cm³ of solution,
   ● calculate the mass (in grams) of solute in 1000 cm³ of solution,
   ● convert the mass (in grams) to moles.

2 To calculate the mass of solute in a certain volume of solution
   ● calculate the mass (in grams) of the solute there is in 1 dm³ of solution,
   ● calculate the mass (in grams) of solute in 1 cm³ of solution,
   ● calculate the mass (in grams) of solute there is in the given volume of the solution.

C3 2.4

How ideas about acids and bases developed

Alkali

The word 'alkali' is derived from Arabic *al qalīy*, meaning the 'calcined ashes'. This refers to the original source of alkalis, which were made from wood ash. The reaction of fats with alkalis produced from wood ash to produce soap (a process called 'saponification') has been known for thousands of years.

Hydrochloric acid

Joseph Priestley discovered hydrogen chloride gas in 1772, when he reacted concentrated sulfuric acid with sodium chloride. When he dissolved the gas in water it produced an acidic solution which became known as muriatic acid. This is from the Latin word *muria*, meaning brine (salt solution).

Joseph Priestley

Lavoisier's ideas

One of the first attempts to explain the behaviour of acids was in 1778, when the French chemist Lavoisier published his ideas about acids, burning and the air. In particular, Lavoisier argued that part of the air was responsible for the behaviour of acids. He called this gas *oxygen*, from the Greek meaning 'acid-forming'.

Lavoisier

Doubts about oxygen and acid

The great British chemist Sir Humphry Davy heated charcoal to high temperatures with HCl and did not get any reaction, and certainly did not see any oxygen produced.

In 1810 Davy wrote:

> One of the singular facts I have observed on this subject, and which I have before referred to, is, that charcoal, even when ignited to whiteness in oxymuriatic or muriatic acid gases, by the Voltaic battery, effects no change in them; if it has been previously freed from hydrogen and moisture by intense ignition in vacuo. This experiment, which I have several times repeated, led me to doubt the existence of oxygen in that substance.

It was not long before Davy concluded that hydrogen, rather than oxygen, is the important element in acids.

Sir Humphry Davy

The important link – acids and hydrogen

It was the German chemist Justus von Liebig who produced the first really useful definition of acids in 1838. He described them as compounds containing hydrogen which could react with a metal to produce hydrogen gas. This definition was not bettered for more than 50 years – although at this time no-one could explain what a base is, other than to describe it as a substance that neutralises an acid.

Justus von Liebig

The breakthrough

The first truly scientific definition of acids and bases was suggested by the Swedish chemist Svante Arrhenius in his PhD thesis in 1884.

Arrhenius suggested that when acids, bases and salts dissolve in water they separate either partly or completely into charged particles called ions in a process called **dissociation**. According to Arrhenius, all the similarities seen between acids with very different formulae were due to the hydrogen ions they produce when dissolved in water. Similarly, he argued that the common properties of bases are due to the fact that they all produce hydroxide ions in solution.

These ideas were seen as revolutionary at the time. The University of Uppsala was very reluctant to give Arrhenius his doctorate, and finally gave him a fourth rank pass – a disgrace which meant he could not get a professorship. Arrhenius kept describing his ideas to other scientists, but older chemists completely rejected his ideas – they were convinced that molecules could not split up and certainly could not carry an electric charge.

Fortunately some of his younger colleagues began to see that Arrhenius's ideas could help explain the results they were getting from their experiments. More and more data built up supporting the new theory and eventually Arrhenius was given credit for the great breakthrough he had made. In 1903, Svante Arrhenius was finally awarded the Nobel Prize for his work.

Svante Arrhenius

The modern way of thinking

Although Arrhenius's theory has been and still is extremely useful, it has some limitations. His definition of acids and bases is limited to situations in which water is present. However, many reactions which appear to be acid–base reactions occur in solvents other than water, when there is no water present at all. For example, think about the reaction between the two gases hydrogen chloride and ammonia. This reaction produces white fumes, made of tiny crystals of ammonium chloride:

$$\text{HCl (g)} + \text{NH}_3\text{ (g)} \rightarrow \text{NH}_4\text{Cl (s)}$$

According to Arrhenius's definition of an acid the reaction between these two gases is not neutralisation because it does not take place in aqueous solution. But it is clearly the same reaction that occurs between HCl and NH_3 in aqueous solution.

This problem was recognised in 1923 by the Danish chemist Johannes Brønsted and the British chemist Thomas Lowry.

Working independently they each produced a much more general definition of acids and bases – that an acid is a **proton donor** and a base is a **proton acceptor**.

This definition explains reactions like the one above as well as aqueous acid–base reactions. The ideas of Lowry and Brønsted were accepted very quickly, because they built on the foundation of Arrhenius's theories which were by now well accepted, and they also helped to explain observations which Arrhenius's ideas could not.

Johannes Brønsted

Thomas Lowry

SUMMARY QUESTIONS

1 Copy the table and draw lines to link the terms with the explanations or definitions.

A	An acid	1	Solution containing a high concentration of acid.
B	An alkali	2	An acid that is partly ionised when dissolved in water.
C	Weak acid	3	Dissolves in water to give $H^+(aq)$ ions.
D	Strong acid	4	Solution containing a low concentration of acid.
E	Concentrated acid	5	Dissolves in water to give $OH^-(aq)$ ions.
F	Dilute acid	6	An acid that is fully ionised when dissolved in water.

2 Not all indicators change colour at the same pH. The table shows four different indicators and the colour changes that occur as the pH increases.

Indicator	Colour change as pH increases	Approximate pH at which colour change occurs
Malachite green	yellow→blue/green	1
Congo red	blue→red	4
Bromothymol blue	yellow→blue	7
Alizarin yellow	yellow→red	11

a) Draw a pH scale which runs from 0 to 14 and draw on it labelled arrows to show where each indicator changes colour.

In the following questions, all solutions have a concentration of 1 mol/dm³.

b) Which indicator could be used to distinguish between a solution containing a strong acid and one containing a weak acid?

c) Which indicator could be used to distinguish between a solution containing a weak acid and one containing a weak alkali?

d) Which indicator could be used to distinguish between a solution containing a strong alkali and one containing a weak alkali?

3 Explain the statement:
'Acids are proton donors whereas bases are proton acceptors.'   [Higher]

EXAM-STYLE QUESTIONS

1 Ethanoic acid is a weak acid. Some students tested solutions of ethanoic acid and hydrochloric acid. The results are shown in the table.

	Hydrochloric acid	Ethanoic acid
Formula	HCl	CH_3CO_2H
Concentration of solution (mol per dm³)	0.10	0.10
pH	1.0	2.9
Reaction with magnesium ribbon	Vigorous bubbling	Slow stream of bubbles
Volume of sodium hydroxide solution used to exactly neutralise 25 cm³ of acid solution (cm³)	25.0	25.0

(a) Explain, as fully as you can, why the pH of ethanoic acid is higher than the pH of hydrochloric acid.   (3)

(b) Explain the difference in the reactions with magnesium ribbon.   (2)

(c) Explain why the volumes of sodium hydroxide to neutralise the acids are the same. You should write balanced equations to help with your explanation. The formula of hydrochloric acid is HCl. Use HA for the formula of ethanoic acid.   (4)

2 A student carried out a titration to find how much dilute nitric acid was needed to react completely with 25 cm³ of sodium hydroxide solution.

Here are his results:

Test	Volume of dilute nitric acid added (cm³)
1	28.9
2	25.1
3	24.9
4	25.0

(a) The student placed 25 cm³ of sodium hydroxide solution into a conical flask to start with. What piece of apparatus should he use to get exactly 25 cm³ of solution? Choose from the list below: (1)

   a beaker   a pipette   a measuring cylinder

(b) What else would be added to the conical flask before adding any acid?   (1)

(c) What piece of apparatus would he use to add the dilute nitric acid to the conical flask? (1)

(d) Which test (1 to 4) produced an anomalous result? (1)

(e) How would the student work out how much acid was needed to neutralise 25 cm³ of dilute sodium hydroxide from his results? (2)

3 25.0 cm³ of a solution of sodium carbonate reacted with exactly 20.0 cm³ of 0.2 mol per dm³ of nitric acid. The equation for the reaction is:

$$Na_2CO_3 + 2HNO_3 \rightarrow 2NaNO_3 + CO_2 + H_2O$$

(a) Name a suitable indicator that could be used to find the end point of this reaction. (1)

(b) Calculate the number of moles of nitric acid in 20.0 cm³ of the solution. (2)

(c) Calculate the number of moles of sodium carbonate in 25.0 cm³ of solution. (1)

(d) Calculate the concentration of sodium carbonate solution in mol per dm³. (2)

(e) Calculate the concentration of the solution in grams per dm³ of anhydrous sodium carbonate. (2)
(A_r values: Na = 23, C = 12, O = 16)

[Higher]

4 25.0 cm³ of a solution containing 4.0 g of sodium hydroxide per dm³ were exactly neutralised by 20 cm³ of a solution of sulfuric acid.

(a) Calculate the concentration of the sodium hydroxide solution in mol per dm³. (2)
(A_r values: Na = 23, O = 16, H = 1)

(b) Calculate the number of moles of sodium hydroxide in 25.0 cm³ of the solution. (2)

(c) Write a balanced equation for the reaction. (2)

(d) Calculate the number of moles of sulfuric acid in 20 cm³ of solution. (1)

(e) Calculate the concentration of the sulfuric acid in mol per dm³. (2)

(f) You have been given 200 cm³ of each of the solutions. Describe how you would carry out a titration to find the volumes of the solutions that react. (6)

[Higher]

HOW SCIENCE WORKS QUESTIONS

You have been asked to design an investigation to find out the changes in pH as you add a strong acid to a strong alkali. You have a burette into which you have been told to add some dilute hydrochloric acid. The flask will contain your sodium hydroxide solution.

You have to add the acid to the alkali and get a reading of the pH as more and more acid is added.

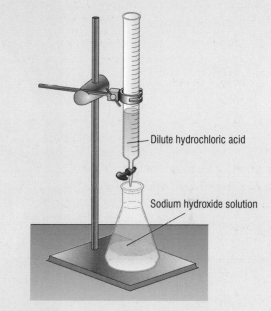

— Dilute hydrochloric acid

— Sodium hydroxide solution

a) You have a choice of either using an indicator and recording the colour changes or a pH meter and recording the pH. Which would you choose? Explain your choice. (1)

b) How much acid would you add at a time? Explain your answer. (2)

c) Why would you stir the flask before checking the pH? (1)

d) Why would it be important to repeat the titration at least one more time? (1)

e) How would you present your results? (1)

f) If you knew that both the acid and the alkali were of the same concentration, how might you check the reliability of your results? (1)

C3 3.1   Water and solubility

Seen from space, the Earth is a blue sphere with green and brown land masses arranged across the surface. The blue colour is water – the most abundant substance on the surface of our planet and essential for all life.

Figure 1 Scientists think that life began in water and remained there for millions of years until plants and animals colonised the land

The water cycle

Water in the rivers, lakes and oceans of the Earth evaporates as the Sun supplies it with energy. The water vapour that forms rises into the atmosphere, where it cools and condenses to form the tiny water droplets that clouds are made from. As the clouds rise further they cool more and the water droplets get bigger. Eventually the droplets fall as rain, replenishing the water in the rivers, lakes and oceans. We call this constant cycling of water between the Earth and the atmosphere – sometimes passing through living organisms on the way – the **water cycle**.

a) What happens to water when it evaporates from the Earth's surface?

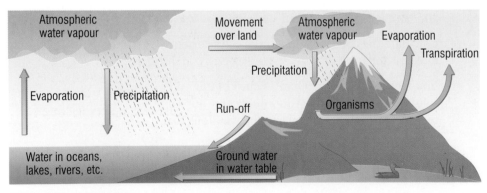

Figure 2 The water cycle

What dissolves?

Many gases are soluble in water to some extent, and so are most ionic compounds. But it is impossible to dissolve many covalent substances in water. In these substances there is little attraction between their molecules and water molecules.

We call the amount of a **solute** which we can dissolve in a certain amount of **solvent** the **solubility** of that substance. We usually measure the solubility of a solute in a solvent (for example, water) in grams of solute per 100 grams of solvent at a particular temperature. The solubility of most solid solutes increases as the temperature increases.

b) What units do we usually use to measure solubility?

A **saturated solution** is one in which as much solute as possible has been dissolved. If we heat the solution, more solute will dissolve until the solution becomes saturated again. As a hot saturated solution cools down some of the solute will come out of the solution – it will crystallise out.

c) What is a saturated solution?

Figure 3 Water shapes our world, surrounds our continents and islands and is essential for life

SUMMARY QUESTIONS

1 Copy and complete using the words below:

> **enlarge   evaporates   rain   saturated   solution**
> **vapour   water cycle**

Water ...... from large masses of water on the Earth's surface and forms .... As this rises water droplets ...... and eventually fall to the surface as .... This circulation of water is called the .... When as much solute as possible is dissolved in a ...... at a particular temperature, we say that the solution is ....

2 Why is it easier to dissolve sugar in hot tea than in cold water?

3 Draw a flowchart to show what happens in the water cycle.

KEY POINTS

1 Water evaporates from rivers, lakes and oceans and condenses to form clouds, returning to the surface as rain.

2 Most ionic substances are soluble in water, but many covalent compounds are not.

3 A saturated solution contains the maximum amount of solute that will dissolve at that temperature.

C3 3.2   Solubility curves

LEARNING OBJECTIVES

1 How does solubility change with temperature?
2 What happens when we cool a saturated solution?
3 Do gases dissolve in water?

Solubility of solids

The amount of a substance that will dissolve in a solvent is affected by the temperature of the solvent. We can see the effect of temperature on the solubility of solutes by looking at graphs which we call **solubility curves**. A solubility curve shows the amount of a solute which dissolves to produce a saturated solution at any given temperature.

a) What does a solubility curve show?

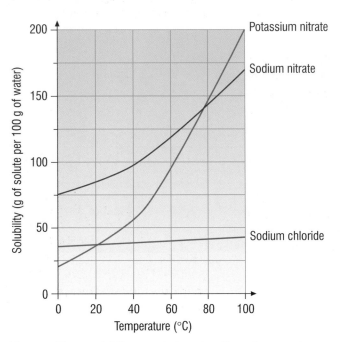

Figure 1 These solubility curves show the effect of temperature on three different solutes dissolving in water. Notice that the solubility of each one increases as the temperature increases.

We can use solubility curves to predict how much solute will form when we cool a hot solution down.

For example, look at the solubility curve of potassium nitrate in Figure 1. You can see that at 80°C, almost 150 g of potassium nitrate will dissolve in every 100 g of water, but at 20°C, only about 35 g will dissolve. That means that as we cool a saturated solution of potassium nitrate down from 80°C to 20°C around 115 g (i.e. 150 g − 35 g) of potassium nitrate will crystallise out of every 100 g of water.

Solubility of gases

There are two main factors which affect the solubility of gases – temperature and pressure. Gases behave in the opposite way to solids when they dissolve in water. As the temperature increases, the amount of gas which will dissolve in a certain volume of water decreases when we keep the pressure constant. On the other hand, if we keep the temperature constant, the solubility of a gas increases as we increase the pressure.

b) What factors affect the solubility of a gas in water?

Gas	Solubility (g of gas in 100 g of water)			
	0°C	20°C	50°C	100°C
Nitrogen (N$_2$)	0.0029	0.0019	0.0012	0
Oxygen (O$_2$)	0.0069	0.0043	0.0027	0
Carbon dioxide (CO$_2$)	0.335	0.169	0.076	0
Ammonia (NH$_3$)	89.9	51.8	28.4	7.4

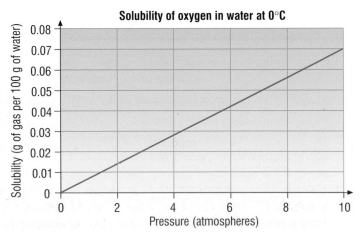

Figure 2 The effect of temperature and pressure on the solubility of gases in water

Why is it important to know about solubility?

There are lots of examples of the way that solubility affects our lives every day. In the kitchen we often need to dissolve solids like sugar in water. This is made easier by heating the water to increase the solubility of the solid.

Outside our homes, the artificial fertilisers that we put on the soil are very soluble in water. They dissolve in rainwater, and can contaminate rivers, lakes and reservoirs.

The nitrate ions in these fertilisers can be very bad for babies and some adults. This means that nitrate levels in drinking water must be carefully monitored by the water companies who supply water to our homes and schools.

Oxygen dissolved in water is essential to keep animals living in water alive. When we pump hot water from the cooling towers of power stations into rivers it contains no chemical pollution.

However, the warm water from the power station increases the temperature of the water in the river. This reduces the amount of oxygen dissolved in it. We call this thermal pollution. It can be very bad for wildlife living in the river, especially fish.

c) Why must drinking water be checked for nitrate levels?
d) Where does nitrate in drinking water come from?

Another example of the importance of dissolved gases is the fizzy drinks that many of us enjoy. The carbonated water we use to make them is produced by dissolving carbon dioxide in water at high pressures before bottling or canning the drink. When we open the can or bottle and release the pressure, the gas comes out of the solution and provides us with 'fizz'!

Lemonade, cola and other sparkling drinks that have been allowed to stand for a while are not very fizzy. That's because they have warmed up and a lot of the carbon dioxide has come out of solution.

Figure 3 Some fish need high levels of oxygen dissolved in the water they live in. They are particularly badly affected if the temperature of the water is raised, reducing the amount of oxygen in solution.

FOUL FACTS

Fizzy drinks make you burp because the carbon dioxide dissolved in them comes out of solution as the drink warms up in your stomach.

SUMMARY QUESTIONS

1 Copy and complete using the words below:

   dissolve   less   more   pressure   solubility   temperature

   The amount of a substance that will …… _dissolve_ in water is shown by a …… _solubility_ curve. The solubility of gases is affected by …… _temp._ and …… _pressure_. Gases are …… _less_ soluble at high temperatures and …… _more_ soluble at high pressures.

2 125 g of sodium nitrate is dissolved in 100 g of water at 80°C. The solution is then allowed to cool to 35°C. Use the graph in Figure 1 to estimate what mass of sodium nitrate crystals will be present at this temperature.   _125 - 90_
   _= 35_

3 Use the information in the table in Figure 2 to plot a graph of the solubility of oxygen in water at temperatures between 0°C and 50°C. Why are fish kept in a small pond in danger of dying in very hot weather?

KEY POINTS

1 The solubility of most solid solutes increases as the temperature rises.
2 The solubility of gases decreases as the temperature rises.
3 Solubility curves show how the solubility of a substance changes with temperature.

C3 3.3   Hard water

LEARNING OBJECTIVES

1 What is hard water?
2 What are the advantages and disadvantages of hard water?

Figure 1 Clean water may look the same wherever you are – but appearances can be deceptive

Most of us in the developed world take it for granted that we have fresh, clean water piped to our homes. The water which comes out of our taps in different areas of the country may look very similar, but there are some very big chemical differences in it. These differences become obvious when we get washed.

When we wash ourselves with soap, the water in some areas of the country forms a really rich, thick lather easily. But in other parts of the country the water doesn't behave in this way. It is quite difficult to get the bubbles that the soap adverts promise us. This is because the water is **hard**.

Figure 2 It isn't always easy to get bubbles like this

Hard water not only makes it difficult to wash ourselves, it also makes it difficult to clean the bath or sink when we have finished. This is because hard water contains dissolved compounds which react with the soap to form **scum**. The scum floats on the water and sticks to the bath.

a) Why is it difficult to wash with hard water?

Most hard water contains dissolved calcium and magnesium compounds. These dissolve when streams and rivers run over or through rocks containing calcium and/or magnesium compounds.

Limestone is an example of such a rock. It contains calcium carbonate. As raindrops fall through the air, carbon dioxide dissolves in them. This dissolved carbon dioxide makes rain slightly acidic, even without pollutants like oxides of sulfur and nitrogen. The water in streams and rivers is therefore slightly acidic too. This means that compounds such as calcium carbonate react and the products formed dissolve in the water.

The dissolved minerals are carried into the reservoirs and on into our domestic water supply. It is the dissolved calcium and magnesium ions that react with soap to form scum.

b) What does hard water form when we put soap in it?

Using hard water is expensive because we need to use much more soap. Before soap ever gets anywhere near dirt it first reacts with the dissolved calcium and magnesium ions in the water, forming salts called stearates (the chemical name for scum). It is only when all of the calcium and magnesium ions have reacted with the soap that a lather can begin to form.

sodium stearate + Ca²⁺ and Mg²⁺ ions → calcium stearate + Na⁺ ions

Figure 3 As scale builds up in heating systems and kettles it not only makes them less efficient – it can stop them working completely

(soap)　calcium and magnesium salts ('hardness')　magnesium stearate precipitate (scum)　soluble in water

As well as forming scum with soap, hard water often leads to **scale** forming in pipes, immersion heaters and other parts of our hot water systems. Pipes can eventually block up. The same scale forms in our kettles, 'furring up' the heating elements and making them much less efficient, because scale is a poor conductor of heat.

c) What does hard water form when we heat it?

$$Ca^{2+}(aq) + 2HCO_3^-(aq) \xrightarrow{\text{heat}} CaCO_3(s) + H_2O(l) + CO_2(g)$$
('hardness')　(limescale)

But hard water isn't all bad news. The same dissolved compounds which are bad for our water pipes seem to be good for our health. Calcium ions in drinking water help in the development of strong bones and teeth. There is also evidence which suggests that hard water helps to reduce the incidence of heart disease in people who drink it.

■ Soft water (Port Arthur)
■ Hard water (Kitchener)

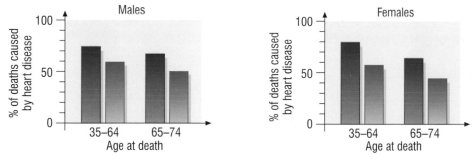

Figure 4 The study on which these graphs are based showed that the number of people suffering from heart disease in a Canadian town with a hard water supply was significantly lower than the number with heart disease in a Canadian town with soft water

SUMMARY QUESTIONS

1 Copy and complete using the words below:

calcium　conductor　efficient　health　magnesium
scale　scum

Hard water contains ...... and/or ...... salts. These react with soap to form .... The salts also produce ...... when the water is heated. This is a poor ...... of heat, and makes kettles and water heaters less .... Hard water may be better for human ...... than soft water.

2 Why is the pH of rainwater less than 7?

3 Using words and chemical equations, explain the difference between *scale* and *scum*.

KEY POINTS

1 Hard water contains dissolved substances such as calcium and magnesium salts.
2 The calcium and/or magnesium ions in hard water react with soap producing a precipitate called scum.
3 The calcium salts and/or magnesium salts also decompose to form scale when the water is heated.
4 Hard water may have benefits for human health.

C3 3.4   Removing hardness

Soft water does not contain dissolved substances that produce scum and scale.

We can soften hard water by removing the calcium and magnesium ions which give it its 'hardness'. Softening water has big benefits for washing ourselves and our clothes, and heating our water too. But people are advised to continue to drink hard water if they can, since scientists think that this is better for them.

Soft water is also important in many industrial processes too, where hardness can produce scale in boilers (making them more expensive to run). Hardness may also interfere with chemical processes like dyeing. There are two important ways to soften hard water.

Figure 1 When we want to dye fabrics we must use soft water – otherwise the fabric may be an uneven colour, or even the wrong colour

a) How is soft water different to hard water?

Method 1 – using washing soda

One way to soften water is to add sodium carbonate to it. Sodium carbonate is also called **washing soda**, because it has been used when people have washed clothes for many years.

When we add washing soda to hard water, it precipitates out calcium and magnesium ions as insoluble carbonates. Once these ions, which cause 'hardness', are no longer in the solution, they cannot react with the soap. This means that the water is soft.

$$Ca^{2+} (aq) + CO_3^{2-} (aq) \rightarrow CaCO_3 (s)$$
('hardness')   (from sodium carbonate)

This reaction is similar to the formation of limescale when hard water is heated. However, here it happens quickly, where and when we want it to happen.

Figure 2 Washing soda is a simple way to soften water without the need for any complicated equipment

b) What is formed when washing soda is used to soften water?

Method 2 – using an ion-exchange column

Water can also be softened by removing the calcium and magnesium ions using an *ion-exchange column*. These columns contain sodium ions which are *exchanged* for the calcium and magnesium ions in hard water when it passes through the column. This is how domestic water softening units work. Some people have these units installed in their houses to soften all of the water used for showering, bathing and washing clothes. A dishwasher contains its own water-softening system.

Once all of the sodium ions in the resin have been exchanged for calcium and magnesium ions, the resin is washed with a salt solution to exchange these ions for sodium ions. This is why water softeners must be kept topped up with salt (sodium chloride). The salt keeps the resin supplied with sodium ions.

c) Why do water softeners need to have salt added to them?

For health reasons, houses fitted with water softeners normally have one cold water tap in the kitchen which is supplied with water which has not been softened.

Figure 3 Water softeners contain ion-exchange resins that enable us to change hard water into soft water

SUMMARY QUESTIONS

1 Copy and complete using the words below:

> calcium   exchange resin   magnesium   scale   scum
> softener   washing soda

Soft water does not contain ...... or ...... ions. This means that it does not produce ...... or .... Hard water can be softened by adding .... , or by using a water ...... which contains an ion- ....

2 Write an equation using words and formulae to show how washing soda softens hard water.

3 Water that has been softened contains sodium ions rather than calcium or magnesium ions. Why does this water not produce scum or scale?

4 Why is it important to have one tap in the kitchen which is not connected to a water softener?

KEY POINTS

1 Soft water does not contain salts that produce scum or scale.
2 Hard water can be softened by removing the salts that produce scum and scale.
3 Water can be softened by adding washing soda or by using an ion-exchange resin to remove calcium and magnesium ions.

241

C3 3.5 Water treatment

LEARNING OBJECTIVES

1 How do we make water safe to drink?
2 What is the difference between pure water and water that is fit to drink?

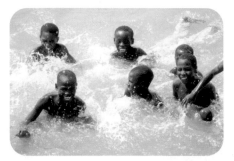

Figure 1 Good, clean water is a precious resource. Those of us lucky enough to have it can too easily take it for granted.

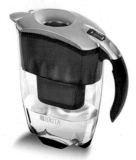

Figure 2 Many people claim that water filters like these improve the taste of water

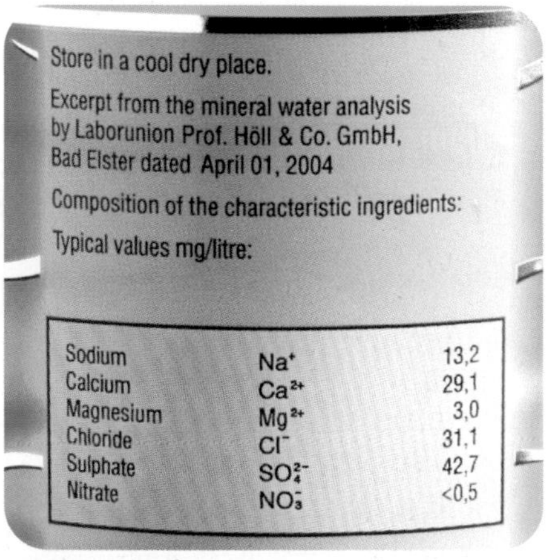

Store in a cool dry place.
Excerpt from the mineral water analysis by Laborunion Prof. Höll & Co. GmbH, Bad Elster dated April 01, 2004
Composition of the characteristic ingredients:
Typical values mg/litre:
Sodium Na^+ 13,2
Calcium Ca^{2+} 29,1
Magnesium Mg^{2+} 3,0
Chloride Cl^- 31,1
Sulphate SO_4^{2-} 42,7
Nitrate NO_3^- <0,5

Figure 3 Mineral waters contain dissolved minerals, which is what gives them their taste and 'character'

Water is a vital and useful resource. It would be impossible to describe all of the ways in which people all around the world use water.

Water is an important raw material, which we use in many industrial processes and as a solvent too. Many substances react with water to form solutions which are either acid or alkaline, and these too have many industrial uses.

Other uses of water worldwide are for washing and cleaning – and of course, for drinking. Providing people with clean drinking water which is uncontaminated by disease, sewage or chemicals, is a major issue all over the world.

Water which comes from boreholes is usually fairly clean because it has been filtered as it travels through the rocks around the borehole. Normally we just need to disinfect water like this with chlorine to make it safe to drink.

When we take water from rivers and reservoirs we usually need to give it more treatment than this. This treatment involves a number of physical and chemical processes. To start with the water source is chosen so that it contains as few dissolved chemicals as possible. The water then passes through five stages. (See Figure 4 on the next page.)

a) How do we normally need to treat water taken from a borehole?
b) Why is chlorine added to water at the end of the treatment process?

Some people use filter jugs in their homes. These usually have a top part into which you put tap water. Figure 2 shows one of these jugs.

As the water goes from the top part of the jug to the lower part it passes through a filter cartridge. This normally contains activated carbon, an ion-exchange resin, and silver.

- The carbon in the filter reduces the levels of chlorine, pesticides and other organic impurities in the water.
- The ion-exchange resin removes calcium, magnesium, lead, copper and aluminium ions.
- Some filter cartridges may contain silver, which discourages the growth of bacteria within the filter.

In most jugs the filter cartridge needs to be changed every few weeks.

Pure – or just fit to drink?

Even water that has been treated and then passed through a jug filter is not pure. It will still contain many substances dissolved in it. But despite this, it is definitely fit to drink. We can get pure water by distilling it – turning impure water into steam and then condensing it – or de-ionising it (using an ion-exchange column to remove all ions except H^+ and OH^-).

We use distilled water in chemistry practical work because it is pure, and so contains nothing but water. But with nothing else in it, distilled water is not very interesting to drink!

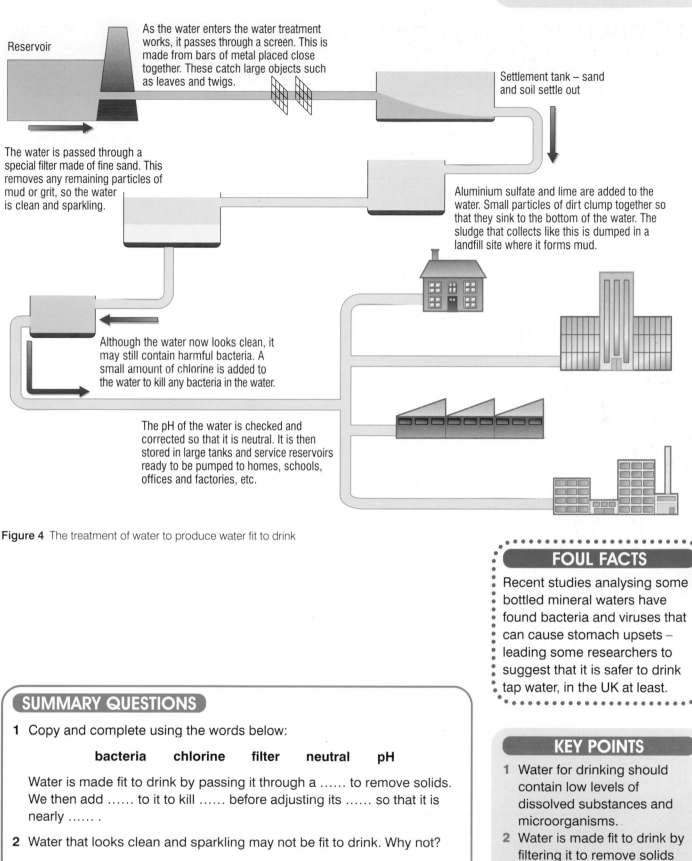

Reservoir

As the water enters the water treatment works, it passes through a screen. This is made from bars of metal placed close together. These catch large objects such as leaves and twigs.

Settlement tank – sand and soil settle out

The water is passed through a special filter made of fine sand. This removes any remaining particles of mud or grit, so the water is clean and sparkling.

Aluminium sulfate and lime are added to the water. Small particles of dirt clump together so that they sink to the bottom of the water. The sludge that collects like this is dumped in a landfill site where it forms mud.

Although the water now looks clean, it may still contain harmful bacteria. A small amount of chlorine is added to the water to kill any bacteria in the water.

The pH of the water is checked and corrected so that it is neutral. It is then stored in large tanks and service reservoirs ready to be pumped to homes, schools, offices and factories, etc.

Figure 4 The treatment of water to produce water fit to drink

SUMMARY QUESTIONS

1 Copy and complete using the words below:

   bacteria   chlorine   filter   neutral   pH

   Water is made fit to drink by passing it through a …… to remove solids. We then add …… to it to kill …… before adjusting its …… so that it is nearly …… .

2 Water that looks clean and sparkling may not be fit to drink. Why not?

3 You could reasonably expect that water, which you have passed through a filter jug, will not form limescale in your kettle. Why?

4 Would it be reasonable to describe bottles of water sold in the supermarket as 'pure'?

KEY POINTS

1 Water for drinking should contain low levels of dissolved substances and microorganisms.

2 Water is made fit to drink by filtering it to remove solids and adding chlorine to kill bacteria.

3 We can make pure water by distillation.

243

C3 3.6   Water fit to drink

'When the water starts boiling it is foolish to turn off the heat.' – Nelson Mandela

The world's poorest people are still waiting for the water itself, let alone for it to boil. The 1980s Water Decade failed to secure water and sanitation for all the world's population. Since then there has been a procession of international reports and conferences calling for universal access to these services.

But constant repetition of the fact that 'water is life' has not proved to be enough. More than a billion people are still without safe water and 2.6 billion lack any way to dispose of their excrement in safety and with dignity.

These failures are undermining development: keeping children out of school, stopping adults pursuing their livelihoods, and denying many people good health and in some cases even life.

This is the silent emergency affecting the world today. Unless the delivery of water and sanitation improves significantly the Millennium Development Goal (MDG) to halve world poverty by 2015 will not be met.

Text extract from the 'Getting to Boiling Point' report, published by Water Aid in March 2005

ACTIVITY

a) i) You are going to carry out a water audit. How much water do you use in a day? The water companies estimate that each of us uses on average 178 litres of water each day! Use the table to work out the water use in **your** home:

	Average use (litres)	Number of times per day in your home	Total water use (litres)
Shower – electric 'power shower'	30 65		
Bath	85		
Flushing the toilet – standard water-saving flush	10 6		
Washing hands and cleaning teeth – under running tap turning tap off	15 10		
Dishwasher	35		
Washing up in sink	10		
Cooking/drinking	10		
Washing machine	80		
Hose (10 minutes to wash car or water garden)	90		

ii) Now imagine you had to fetch all of this water in 20 litre pots.
What would be the mass of all this water? (1 litre of water = 1 kg)
How many pots would be needed?
Estimate how long it might take if the water supply was 5 km away from your home. Could you use this much water each day?

b) Why are women and children particularly affected by poor sanitation and water supplies? Design a poster as part of a campaign to draw attention to the need to bring clean water and sanitation to villages in the developing world. Use your poster to explain how this will help children especially.

c) Write a letter that you could send to the editor of your local paper, to your MP or to anyone else that you think you might write to about the need for clean water and sanitation in the developing world.
Make your points clearly, and back up your arguments with facts.

SUMMARY QUESTIONS

1 Link the statements a) to c) with options from A to E.

a) Water evaporates from lakes, rivers and oceans, forms clouds and then falls as rain.	A Cause hardness in water.
b) Calcium and magnesium ions.	B Calcium salts.
c) Causes limescale when water is boiled.	C This is the **water cycle**.
	D React with soap to form scum.
	E Exchanged for sodium or potassium ions in a water softener.

2 A student carried out an experiment to measure the solubility of two different substances in water at different temperatures. One of these substances was a crystalline solid, the other was a gas. The two graphs A and B show her results.

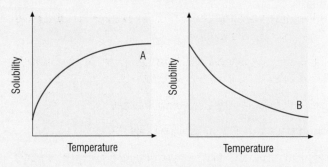

a) Which graph, A or B, shows the results obtained by the student for the solubility of the gas? Explain your answer.

b) Explain the shape of graph A.

c) Warm water from a power station's cooling system is discharged into a river. Use graph B to explain why there are fewer fish in this part of the river than elsewhere along its length.

3 Washing soda is used to soften water, making it easier to get clothes clean. It consists of crystals of sodium carbonate.

a) Name **one** ion which makes water hard.

b) Write a balanced equation for the reaction of washing soda with this ion, showing how the washing soda makes the water soft.

EXAM-STYLE QUESTIONS

1 Fizzy drinks are made using carbonated water.

(a) Name the gas that is released when a can of fizzy drink is opened. (1)

(b) Explain why bubbles form in the drink when it has been poured into a glass. (2)

(c) Why does cooling the can before it is opened reduce the amount of bubbles that are produced? (1)

2 The solubility of potassium chlorate in water at different temperatures is shown in the table.

Temperature (°C)	10	20	30	50	60	80	90
Solubility (g per 100 g water)	5	7	10	19	24	37	46

(a) i) Plot a graph of these results. Put temperature on the horizontal axis and solubility on the vertical axis. Draw a smooth line through the points. (4)

ii) What do we call the smooth line drawn through points obtained by experiment? (1)

(b) What does the graph show about how the solubility of potassium chlorate changes with temperature? (1)

(c) Use your graph to find the solubility of potassium chlorate at 70°C. (1)

(d) Explain what is meant by a saturated solution. (2)

(e) A saturated solution of potassium chlorate was made using 100 g of water at 55°C. What mass of potassium chlorate separates when the solution is cooled to 25°C? (2)

(f) A student wanted to check the solubility of potassium chlorate at 20°C by experiment. The value given in his data book was 7 g per 100 g of water. He did the experiment 4 times. His results are shown below:

Number of experiment	Solubility (g per 100 g of water)
1	5
2	9
3	8
4	6

i)   What was the range of his results presented in the table? (2)

ii) Work out the mean (average) value for the solubility of potassium chlorate at 20°C from his results (1)

iii) Comment on the accuracy of the student's results. (2)

iv) The first time he did the test his result came out at 15 g per 100 g of water. He did not include this in his results table. What is this type of result called? (1)

3   Water from a spring was found to be hard. A sample of 1.0 dm³ of the water was evaporated. The solid residue was analysed and the results are shown in the table.

Name of mineral	Amount (mg)
Calcium carbonate	86
Magnesium sulfate	39
Magnesium chloride	34
Sodium chloride	29
Sodium nitrate	5

(a) Name all the compounds in the table that make this water hard. (2)

(b) Explain why a scum is formed when soap is used with hard water. (2)

(c) (i) Suggest one method that could be used to make this water soft. (1)

    (ii) Explain how this method works. (1)

(d) The calcium carbonate in the residue reacts with hydrochloric acid.

    (i) Write a balanced equation for this reaction. (2)

    (ii) What volume of hydrochloric acid, concentration 0.10 mol per dm³, would react with the calcium carbonate in the residue from 1 dm³ of this water? (3)

(e) Explain why hot water pipes and central heating boilers should be checked regularly in hard water areas. (2)

[Higher]

HOW SCIENCE WORKS QUESTIONS

Yasmin carried out an investigation into the solubility of potassium nitrate. She carefully weighed a beaker. She added 100 cm³ of pure water. She re-weighed the beaker. She then carefully added her potassium nitrate, stirring after adding each amount, until she could see a sediment in the beaker. She then re-weighed the beaker.

She did this with different temperatures of water. Her results were as follows:

At 26°C she added 43 g of potassium nitrate. At 44°C she added 74 g and at 65°C she added 128 g.

a) Produce a table of Yasmin's results. (3)

b) Draw a graph of Yasmin's results. (3)

c) Do you think Yasmin produced enough results? Explain your answer. (1)

d) The work took a very long time and Yasmin did the work very carefully. Should she have done any repeats? Explain your answer. (1)

e) i) Yasmin checked her results against a table of saturation temperatures for potassium nitrate in a data book.
       Why would she do this? (1)

   ii) Yasmin was quite pleased with her results. The values in the data book were much the same as her own results, except that hers were all slightly higher than the ones in the book.
       Why might this have happened? (2)

C3 4.1

Comparing the energy produced by fuels

LEARNING OBJECTIVES

1 How can we measure the energy produced by different fuels?

2 What types of food produce most energy?

We have already seen on page 174 that exothermic reactions produce energy. When we burn a fuel we use an exothermic reaction as a source of energy. This may be to keep ourselves warm (when we burn oil or wood, for example) or it may be to get ourselves or things we want from one place to another.

a) Why do we use fuels?

Figure 1 Keeping warm or moving about – we need exothermic reactions

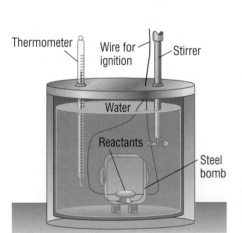

Figure 2 A bomb calorimeter is used to make accurate measurements of energy changes in reactions. The energy change in the reaction causes the temperature of the water in the calorimeter to change. The energy change is then calculated using this temperature change and the mass of the water in the calorimeter.

Not all fuels produce the same amount of energy when they burn – some reactions are more exothermic than others. It is often very important to know how much energy a fuel produces when it burns. Measurements like this are carried out under carefully controlled conditions, in an instrument called a **bomb calorimeter**. (See Figure 2 opposite.)

b) What instrument do we use to measure how much energy is produced when a fuel burns?

In a school chemistry lab it is very difficult to measure accurately the energy produced by fuels when they burn. But we can use simple methods to **compare** the energy produced by different fuels.

c) What variables must we control in order to compare the energy released by different fuels?

Food is a fuel too!

The amount of energy in food is measured in exactly the same way. There are data tables that we use to work out which food to eat if we are on a diet. They are drawn up by burning the foods in a bomb calorimeter and measuring the energy that is produced. Measurements like this show how different types of food produce very different amounts of energy when they react with oxygen.

Just as we can compare the energy that fuels produce by using a simple calorimeter, we can compare the energy content of foods – providing they burn.

Food	Energy per 100 g	Food	Energy per 100 g
bread	1120 kJ	apple	160 kJ
chocolate biscuit	2240 kJ	celery	32 kJ
butter	3310 kJ	sausages (fried)	1450 kJ

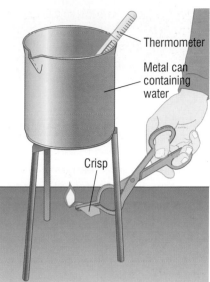

Figure 3 The energy released by foods when they burn can be compared using some very simple equipment

PRACTICAL

Comparing the energy produced by fuels when they burn

To compare the energy produced by different fuels when they burn, the fuels are used to heat up water in a copper can or a glass beaker. By measuring the temperature changes produced by different fuels we can compare the energy they release when they burn.

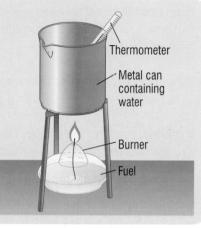

GET IT RIGHT!

Measurements using simple calorimeters are not accurate because of heat losses, but they can be used to compare the amounts of heat produced.

SUMMARY QUESTIONS

1 Copy and complete using the words below:

   calorimeter   respire   energy   exothermic   oxygen

   When fuels burn they react with .... This reaction releases ...... – it is .... Food also releases energy when we .... The energy content of foods and fuels can be measured using a ....

2 Look at the table above showing different foods. What kind of foods produce
   a) most energy,
   b) least energy when we eat them?

3 A simple calorimeter was used to compare the energy produced by three different fuels. The results were as follows:

	Mass of fuel (g)	Volume of water (cm³)	Temperature change (°C)
Fuel A	0.24	160	10.0
Fuel B	0.18	100	8.0
Fuel C	0.27	150	9.0

   a) For fuels A, B and C, calculate the mass of each fuel required to increase the temperature of the water by 1°C.
   b) Now calculate the mass of fuel required to increase the temperature of 1 cm³ of water by 1°C.
   c) Arrange the fuels in order of the amount of energy they produce.

KEY POINTS

1 When fuels and food react with oxygen, energy is released (the reaction is exothermic).
2 A simple calorimeter can be used to compare the energy released by different fuels or different foods in a school chemistry lab.

C3 4.2 Energy changes in reactions

1 What happens to the bonds in a chemical reaction?
2 How do we represent energy changes in chemical reactions?
3 What do we mean by activation energy?

When we carry out a chemical reaction, we can think of the chemical bonds between the atoms in the reactants being broken. Then new chemical bonds can be formed to make the products.

- Because energy has to be supplied to break chemical bonds, breaking bonds is an **endothermic** process – energy is taken in from the surroundings.
- But when new bonds are formed, energy is released – so making bonds is an **exothermic** process.

a) What kind of process is breaking bonds?
b) What kind of process is making bonds?

- In some reactions the energy released when new bonds are formed (as the products of the reaction are made) is more than the energy needed to break the bonds in the reactants. These reactions transfer energy to the surroundings – they are **exothermic**.

- In other reactions the energy needed to break the bonds in the reactants is more than the energy released when new bonds are formed in the products. These reactions involve a transfer of energy from the surroundings to the reacting chemicals – they are **endothermic**.

c) If the energy required to break bonds is greater than the energy released when bonds are made, what kind of reaction is this?

The balance between the energy needed to break bonds and the energy released when new bonds are made is what decides whether a reaction is endothermic or exothermic. We can find out more about what is happening in a particular reaction by looking at energy level diagrams.

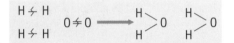

Figure 1 When hydrogen and oxygen react to make water, bonds between hydrogen atoms and between oxygen atoms have to be broken before bonds between oxygen atoms and hydrogen atoms can be formed

Remember that **B**reaking bonds a**B**sorbs energy, fo**R**ming bonds **R**eleases energy.

Energy level diagrams

Energy level diagrams show us the relative amounts of energy contained in the reactants and the products of a reaction. This energy is measured in **kJ/mol**.

Figure 2 shows the energy level diagram for an exothermic reaction. The products are at a lower energy level than the reactants, so as reactants react to form products energy is released.

The difference between the energy levels of the reactants and the products is the change during the reaction, measured in kJ/mol. We represent this energy change by the symbol ΔH ('delta H'). This simply means 'the difference in energy between the products and the reactants'.

- In an exothermic reaction ΔH is always negative, because the products are at a lower energy level than the reactants. The temperature of the surroundings increases.

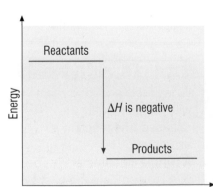

Figure 2 An exothermic reaction – more energy is released when bonds are formed between atoms in the products than is needed to break the bonds between the atoms in the reactants

Figure 3 on the next page shows the energy level diagram for an endothermic reaction. Here, the products are at a higher energy level than the reactants. As reactants react to form products, energy is absorbed from the surroundings.

Once again, the difference between the energy levels of the reactants and the products is the energy change during the reaction.

- In an endothermic reaction ΔH is always positive, because the products are at a higher energy level than the reactants. The temperature of the surroundings decreases.

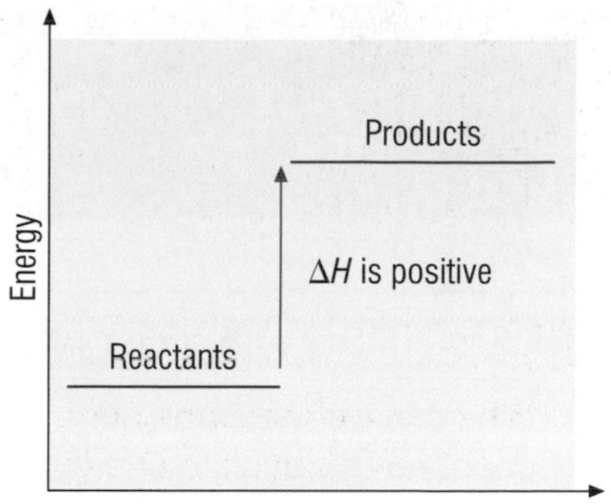

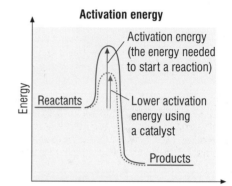

Figure 3 An endothermic reaction – more energy is needed to break the bonds between the atoms in the reactants than is released when bonds are formed between atoms in the products

---

Worked example

We can use a simple calorimeter like the one on page 175 to measure the energy change in a chemical reaction. To do this we use the fact that:

4.2 J of energy raise the temperature of 1 g of water by 1°C

A simple calorimeter is used to measure the energy change in the reaction:

$$A + B \rightarrow C$$

60 cm³ of a solution containing 0.1 moles of A is mixed with 40 cm³ of a solution containing 0.1 moles of B. The temperature of the two solutions before mixing is 19.6°C – after mixing them, the maximum temperature reached is 26.1°C.

Step 1 – calculate temperature change:

$$\text{temperature change} = 26.1°C - 19.6°C$$
$$= 6.5°C$$

Step 2 – calculate energy change:

4.2 J of energy raise the temperature of 1 g of solution by 1°C:

[mass solution = 100 g (assuming that density of solution = density of water, which is true for dilute solutions)]

$$\text{energy change} = 100\,g \times 4.2\,J/g/°C \times 6.5°C$$
$$= 2730\,J$$
$$= 2.73\,kJ$$

This is the energy change when 0.1 moles of reactants are mixed – so when 1 mole of reactants are mixed the energy change will be:

$$= 2.73\,kJ \times 10$$
$$= 27.3\,kJ$$

So this experiment gives the energy change for the reaction:

$$A + B \rightarrow C$$

as **−27.3 kJ/mol**. (The temperature rises so the reaction is exothermic.)

Activation energy

Figure 4 The energy needed to start a reaction is called the **activation energy**. A catalyst lowers the activation energy so that a higher proportion of reactant particles have sufficient energy to react.

SUMMARY QUESTIONS

1 Copy and complete using the words below:

   absorbing   formed   positive   released   releases   endothermic

   Breaking bonds involves …… energy, while making bonds …… energy. If more energy is needed to break bonds than is …… when bonds are ……, the reaction is …… and ΔH is …… .

2 Why is energy released in an exothermic reaction?

3 Draw energy level diagrams for the following reactions:

   a) $6CO_2\,(g) + 6H_2O\,(g) \rightarrow C_6H_{12}O_6\,(aq) + 6O_2\,(g)$; $\Delta H = +2880\,kJ/mol$
   b) $H_2\,(g) + I_2\,(g) \rightarrow 2HI\,(g)$; $\Delta H = +26.5\,kJ/mol$

KEY POINTS

1 In chemical reactions, energy must be supplied to break the bonds between atoms in the reactants.

2 When bonds are formed between atoms in a chemical reaction, energy is released.

3 In an exothermic reaction, the energy released when bonds are formed is greater than the energy absorbed when bonds are broken. The opposite is true for endothermic reactions.

4 ΔH is negative for exothermic reactions. It is positive for endothermic reactions.

5 The minimum amount of energy to start a reaction is called the activation energy.

C3 4.3   Calculations using bond energies

LEARNING OBJECTIVE

1 How can we use bond energies to calculate energy changes in reactions?

The energy needed to break the bond between two atoms is known as the **bond energy** for that bond.

Bond energies are measured in kJ/mol. We can use bond energies to work out the energy change (ΔH) for many chemical reaction. Before we can do this, we need to have a list of the bond energies for some of the most commonly found chemical bonds:

Bond	Bond energy (kJ/mol)	Bond	Bond energy (kJ/mol)
C—C	347	H—Cl	432
C—O	358	H—O	464
C—H	413	H—N	391
C—N	286	H—H	436
C—Cl	346	O=O	498
Cl—Cl	243	N≡N	945

To calculate the energy change for a chemical reaction we need to work out:

● firstly, how much energy is needed to break the chemical bonds in the reactants, and
● then how much energy is released when the new bonds are formed in the products.

a) What do we mean by the **bond energy** of a chemical bond?

It is very important to remember that the data in the table is the energy required for **breaking** bonds. When we want to know the amount of energy released as these same bonds are formed, the number is the **same** but the sign is **negative**.

For example, the bond energy for a C—C bond is +347 kJ/mol. This means that the bond energy for **forming** a C—C bond is −347 kJ/mol.

b) Is bond making endothermic or exothermic? What about bond breaking?

DID YOU KNOW?

Bond energies in different molecules are remarkably similar, which is why it is possible to use them to calculate energy changes for reactions in this way. However, values do vary slightly depending on the molecule so ΔH worked out like this gives us an approximate value.

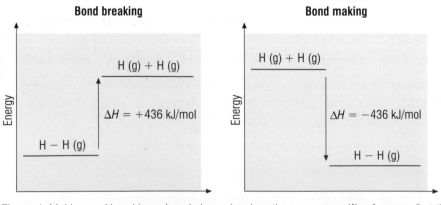

Figure 1 Making and breaking a bond always involves the same **quantity** of energy (but the **sign** is different)

Worked example

Ammonia is made from nitrogen and hydrogen in the Haber process. The balanced chemical equation for this reaction is:

$$N_2 (g) + 3H_2 (g) \rightarrow 2NH_3 (g)$$

Calculate the overall energy change in this reaction.

Solution

This equation tells us that we need to break apart 1 mole of nitrogen molecules and 3 moles of hydrogen molecules in the reaction.

Nitrogen molecules are held together by a triple bond (written like this, N≡N). This bond is very strong – from the table, its bond energy is 945 kJ/mol.

Hydrogen molecules are held together by a single bond (written like this, H—H). From the table, we can see that the bond energy for this bond is 436 kJ/mol.

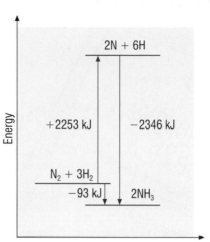

Figure 2 The formation of ammonia. The energy change of −93 kJ is for the formation of *two* moles of ammonia. So if you wanted to know the energy change for the reaction *per mole of ammonia* formed, it would be exactly half this, i.e. −46.5 kJ/mol.

Energy needed to break 1 mole of N≡N and 3 moles of H—H bonds
= 945 + (3 × 436) kJ = **+2253 kJ**

When these atoms form ammonia (NH₃), 6 N—H bonds are made as 2 moles of NH₃ are formed. The bond energy of the N—H bond is 391 kJ/mol.

Energy released when 6 moles of N—H bonds are made = 6 × −391 kJ
= **−2346 kJ**

Figure 2 shows the overall energy change in this reaction = +2253 − 2346 kJ = **−93 kJ**

SUMMARY QUESTIONS

1 Nitrogen (N_2) is a very unreactive element. Why?

2 Write balanced equations and calculate the energy changes for the following chemical reactions:
(Use the bond energies supplied in the table on page 252.)

a) hydrogen + chlorine → hydrogen chloride
b) oxygen + hydrogen → water

KEY POINT

1 The overall energy change in a chemical reaction can be calculated using bond energies.

C3 4.4   Energy balance – how much energy do you use?

Energy from fuels ...

How Much Energy do YOU use?!

Please answer the following questions about your lifestyle:

1. How far is it from your home to your school in kilometres?

2. How do you get to school (choose one):
(a) walk (b) cycle (c) bus (d) car

If (a) or (b) – well done! The energy resources you use for your journey to school mean that your impact on the environment is about as low as it could be!

If (c) – your energy use isn't as good as (a) or (b), but it's still better than (d). Well done!

If (d) – oh dear! Could you cycle or walk? Should you move to be nearer the school?

3. Now let's work out how much energy you use.

ACTIVITY

Although it can seem simple, reducing the amount of energy that we use isn't always easy since much of the way we live is based on the assumption that we have access to energy resources. For example, many children travel long distances to school. For some this may be because they have made a choice about where they want to go to school. But others may have little choice.

Working in groups, plan and write the script for a television or radio programme looking at the issues concerned with the amount of energy we use to travel to school and work. Try to look at both sides of the picture:

● on one hand, those of us who live in the developed world should reduce our share of the world's energy resources that we use;
● on the other hand, it's not always easy to live close to where you go to school or work, and sometimes we have no practical alternative to travelling by car or bus.

Energy from food ...

Child in developing world

Time spent:
- ☐ Fetching water
- ☐ Doing chores at home
- ☐ Working in the fields
- ☐ Preparing, cooking and eating food
- ☐ School
- ☐ Collecting firewood for cooking
- ☐ Sleeping

Child in developed world

Time spent:
- ☐ Watching TV
- ☐ Playing computer games
- ☐ School
- ☐ Sleeping

About a quarter of the time spent watching TV or playing computer games is spent eating!

ACTIVITY

Where do you go if you're hungry? The larder? The kitchen cupboards? And if they're empty – to the corner shop, or the supermarket?

In the developed world we don't usually find it hard to get food. Most of us don't have to spend our days worrying about how we'll make the next meal, or spend hours collecting the food to make it, or the fuel to cook it. But it isn't the same in the developing world – where many people spend a lot of time and energy just finding the food, water and fuel they need to live.

The two pie charts show how two different children spend their days. They are both lucky – they have enough to eat and a place to sleep.

Write a short feature article for a newspaper or a news website comparing the lives of these two children. In your article, explain why the lives of children in the developed world make it more likely that they will be overweight than children in the developing world. This is true even when we compare children who both have similar amounts to eat.

SUMMARY QUESTIONS

1 Match statements a) to d) with the terms labelled A to D in the table below.

a)	Reactant required for fuels and food to release energy.	A	bomb calorimeter
b)	Reaction in which energy is released.	B	calorie
c)	Apparatus used to measure energy change in a reaction.	C	exothermic
d)	Unit used to describe the energy released when food is eaten.	D	oxygen

2 When we eat sugar we break it down to produce water and carbon dioxide.

$$C_{12}H_{22}O_{11} + 12O_2 \rightarrow 12CO_2 + 11H_2O$$

a) Why must your body *supply* energy in order to break down a sugar molecule?

b) When we break down sugar in our bodies, energy is released. Explain where this energy comes from in terms of the bonds in molecules.

c) We can get about 1.7×10^3 kJ of energy by breaking down 100 g of sugar.

If a heaped teaspoon contains 5 g of sugar, how much energy does this produce when broken down by the body?

3 Hydrogen peroxide has the structure H—O—O—H. It decomposes slowly to form water and oxygen.

$$2H_2O_2 \rightarrow 2H_2O + O_2$$

The table shows the bond energies for different types of bond.

Bond	Bond energy (kJ/mol)
H—O	464
H—H	436
O—O	144
O=O	498

a) Use the bond energies to calculate the energy change for the decomposition of hydrogen peroxide in kJ/mol.

b) Is this reaction exothermic or endothermic? Explain your answer. [Higher]

EXAM-STYLE QUESTIONS

1 A student burned some liquid fuels using the apparatus shown in the diagram.

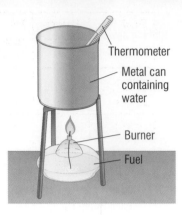

The student used the same volume of water and the same starting temperature of the water for each fuel. The student's results are shown in the table.

Fuel	Mass of fuel burned (g)	Temperature rise of water (°C)	Temperature rise per gram of fuel burned (°C/g)
Ethanol	2.46	32	13.0
Kerosene	1.30	30	
Methanol	2.40	25	
Petrol	1.80	38	

(a) Calculate the values of the temperature rise per gram of fuel burned that are missing from the table. (3)

(b) Write down the four fuels in order of increasing heat energy produced per gram of fuel according to the experiment. (1)

(c) The student checked the results with some data for fuels in a textbook. The values from the experiment were much lower than the values in the book.

  (i)   Suggest one reason why this experiment gives lower results than expected. (1)

  (ii) What sort of error is this? (1)

(d) Suggest one way that the student could change this experiment to make it a fairer comparison. Explain why your suggestion would make it fairer. (2)

2 Two students had snacks. The first table shows what was in each student's snack. The second table gives some nutritional information about the foods.

Student 1	Student 2
100 g French fries	25 g crisps
300 g cola drink	250 g fruit smoothie

Food	Energy (kJ per 100g)	Fat (g per 100g)	Protein (g per 100g)	Carbohydrate (g per 100g)
Cola	176	0	0	11
Crisps	2216	34	5.7	53
French fries	1174	15	3.3	34
Fruit smoothie	194	0.5	1.5	20

(a) Calculate the total energy produced by each student's snack. (4)

(b) Student 1's snack produces more energy than the other snack. Suggest **two** reasons why this snack produces more energy. (2)

(c) What is the effect of eating food that provides more energy than the body needs? (1)

3   Methane is used as a fuel. The equation for the complete combustion of methane is:

$$CH_4 + 2O_2 \rightarrow CO_2 + 2H_2O$$

The energy changes for the reaction are shown on the diagram:

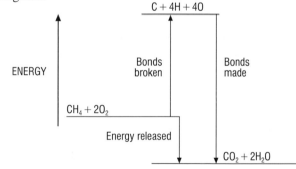

Bond	Bond energy (kJ/mol)	Bond	Bond energy (kJ/mol)
C—H	413	C=O	805
O=O	498	H—O	464

(a) Draw the structure of methane to show all the bonds in methane. (2)

(b) Use the table of bond energies and the energy diagram to calculate:

  (i) the energy to break all the bonds in the reactants. (2)

  (ii) the energy produced when the bonds in the products are made. (2)

  (iii) the energy released by the reaction. (1)

[Higher]

HOW SCIENCE WORKS QUESTIONS

Bomb calorimeter

Bomb calorimeters are used in research laboratories to calculate the energy in food. You will have tried to do this at some time in school when you burn a crisp or some fuel under a beaker of water and measure the rise in temperature. A bomb calorimeter does the same thing, but better!

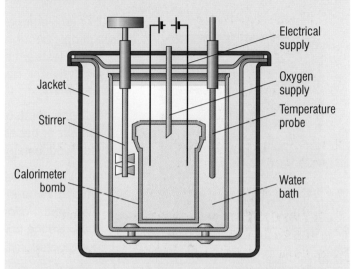

As you can see from this diagram, it is sealed and has its own supply of oxygen. There is a supply of electricity to produce a spark that will ignite the food. The increase in temperature of the water around the bomb is measured.

a) Why is the bomb 'better' than the school method? (1)

b) The bomb is calibrated before it is used by burning a substance with a known energy value. Use this example to explain what is meant by 'calibration'. (1)

Most foods are not tested in a bomb calorimeter. They have their energy value calculated by what is called the 4–9–4 rule. That is, for every gram of protein there are 4 calories of energy; 9 for fats and 4 for carbohydrates. So once you know the content of the food you can simply calculate its energy value.

c) Will the 4–9–4 rule or the bomb calorimeter produce the most valid results? Explain your answer. (1)

d) The bomb calorimeter burns parts of the food that are not digestible, e.g. hair which is protein but cannot be digested. How does this affect your answer to c)? (1)

C3 5.1   Tests for positive ions

1 How do we identify different positive ions?

Figure 1 A flame test can identify a Group 1 or Group 2 metal in a compound. In this case the metal is lithium

Element	Flame colour
lithium	bright red
sodium	golden yellow
potassium	lilac
calcium	brick red
barium	green

One of the problems we face in chemistry is that many substances look very similar to the naked eye. Just telling the difference between elements is hard enough. There are lots of shiny grey metals and one colourless, odourless gas looks much like another!

But once elements have been combined to form compounds the possibilities are simply enormous. To identify unknown substances we have a variety of different tests which can help us to distinguish one substance from another.

Flame tests

Identifying some of the metals in Groups 1 and 2 of the periodic table is made much easier because most of them produce flames with a characteristic colour. To carry out a flame test you:

- put a small amount of the compound to be tested in a platinum wire loop which has been dipped in concentrated hydrochloric acid,
- then hold the loop in the roaring blue flame of a Bunsen burner,
- then use the colour of the Bunsen flame to identify the metal element in the compound.

a) What do we use flame tests for?

Sometimes we can use the reactions of unknown compounds with **sodium hydroxide solution** to help us with our identification. Aluminium ions, calcium ions and magnesium ions all form **white precipitates** with sodium hydroxide solution. So if we add sodium hydroxide to an unknown compound and a white precipitate forms we know it contains either aluminium, calcium or magnesium ions.

If we add more and more sodium hydroxide then the precipitate formed with aluminium ions dissolves – but a precipitate formed with calcium or magnesium ions will not. Calcium and magnesium ions can be distinguished by a flame test. Calcium ions give a brick red flame but magnesium ions produce no colour at all.

b) What colour precipitate does sodium hydroxide produce with aluminium, calcium and magnesium ions?

Some metal ions form **coloured precipitates** with sodium hydroxide.

- If we add sodium hydroxide solution to a substance containing copper(II) ions a light blue precipitate appears.
- If the substance contains iron(II) ions a 'dirty' green precipitate is produced when sodium hydroxide solution is added.
- When sodium hydroxide solution is added to iron(III) ions, a reddish-brown precipitate is produced.

Figure 2 This distinctive precipitate formed when we add sodium hydroxide solution tells us that Cu^{2+} ions are present

Sodium hydroxide solution can also be used to test whether ammonium ions (NH_4^+) are present in an unknown substance. Ammonium ions react with sodium hydroxide solution to form ammonia and water:

$$NH_4^+ \text{ (aq)} + OH^- \text{ (aq)} \rightarrow NH_3 \text{ (aq)} + H_2O \text{ (l)}$$

To test for ammonium ions, we add sodium hydroxide solution to a solution of the unknown substance. If ammonium ions are present, ammonia is formed. When we gently warm the solution, ammonia is then driven off as a gas. We can detect this using damp red litmus paper. The red litmus turns blue because ammonia is an alkaline gas.

PRACTICAL

Identifying positive ions

Try to identify the metal in some unknown compounds.

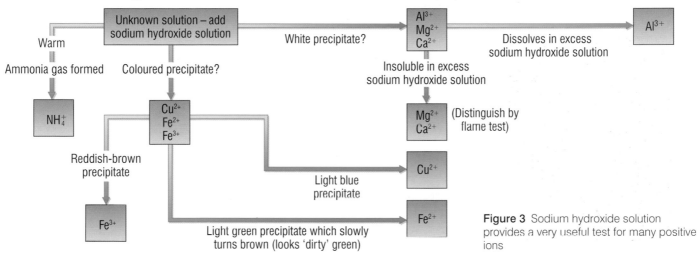

Figure 3 Sodium hydroxide solution provides a very useful test for many positive ions

---

SUMMARY QUESTIONS

1 Copy and complete using the words below:

   **ammonia   ammonium   colours   flame   hydroxide   one
   precipitates   two**

   The Group …… and …… elements produce distinctive …… when their compounds are placed in a …… . Sodium …… solution can be used to identify metal ions by the …… that they produce. It can also be used to identify …… ions, which produce …… gas when warmed.

2 Draw a flowchart to describe how to carry out a flame test.

3 Copy and complete the table.

Add sodium hydroxide solution	Flame test	Metal ion
nothing observed	lilac	e)
white precipitate	brick red	f)
a)	c)	Fe^{3+}
white precipitate which dissolves as more sodium hydroxide solution added	nothing observed	g)
light green precipitate which slowly turns reddish brown	nothing observed	h)
b)	d)	Na^+

DID YOU KNOW?

Many metals are poisonous in large enough amounts – so it is very important to be able to analyse blood and body tissues for metal ions.

KEY POINTS

1 Group 1 and Group 2 metals can be identified in their compounds using flame tests.

2 Sodium hydroxide solution can be used to identify different metal ions depending on the precipitate that is formed.

3 Ammonium ions produce ammonia when sodium hydroxide solution is added and the solution is warmed gently.

C3 5.2

Tests for negative ions

Carbonates

If we add a dilute acid (such as hydrochloric acid) to a carbonate it fizzes and produces carbon dioxide gas. This is a good test to see if an unknown substance is a carbonate.

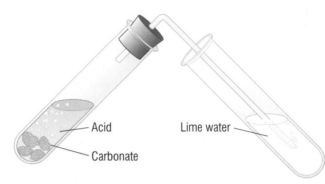

Acid

Lime water

Carbonate

Figure 1 The test for carbonates

Two particular metal carbonates have very distinctive colour changes when we heat them. This means that if we have identified an unknown substance as a carbonate, and then heat some of that substance, we may be able to complete the identification. For example, copper carbonate is a green substance which decomposes when we heat it to give black copper oxide and carbon dioxide:

$$CuCO_3 \text{ (s)} \xrightarrow{\text{heat}} CuO \text{ (s)} + CO_2 \text{ (g)}$$
green   black   turns lime water milky

Similarly, zinc carbonate is a white substance which decomposes to give zinc oxide when heated. Zinc oxide is also white, but when it is hot it is a lemon-yellow colour, turning back to white as it cools down. It is the only substance to show such a colour change:

$$ZnCO_3 \text{ (s)} \xrightarrow{\text{heat}} ZnO \text{ (s)} + CO_2 \text{ (g)}$$
white   lemon-yellow when hot,   turns lime water milky
                white when cold

a) What gas is produced when we add dilute acid to a carbonate?

Halides (chloride, bromide and iodide)

A very simple test shows whether chloride, bromide or iodide ions are present in a compound. If we add dilute nitric acid and silver nitrate solution to the unknown solution, the appearance of a precipitate tells us that one of the halide ions is present.

The colour of the precipitate tells us which halide it is:

- chloride ions give a white precipitate
- bromide ions give a cream precipitate, and
- iodide ions give a pale yellow precipitate.

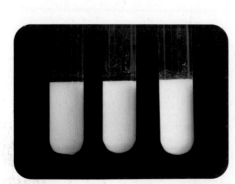

Figure 2 One simple test can tell us if an unknown substance contains chloride, bromide or iodide ions

Here is the ionic equation, where X⁻ is the halide ion:

$$Ag^+ (aq) + X^- (aq) \rightarrow AgX (s)$$

b) How do we test for halide ions?

Sulfates

Sulfate ions in solution produce a white precipitate when we add hydrochloric acid to them followed by barium chloride solution. The white precipitate is the insoluble salt, barium sulfate.

Here is the ionic equation:

$$Ba^{2+} (aq) + SO_4^{2-} (aq) \rightarrow BaSO_4 (s)$$

c) How do we test for sulfate ions?

Nitrates

To detect the presence of nitrate ions in an unknown compound we make use of the test for ammonia that we saw on page 259.

Once again, we add sodium hydroxide solution to a solution of the unknown substance and gently warm it. If no ammonia is detected, we add a little aluminium powder. The aluminium powder reduces the nitrate ions to ammonium ions. These then react with the sodium hydroxide solution to form ammonia gas, which is given off. This is detected using damp red litmus paper which turns blue.

d) How do we test for nitrate ions?

SUMMARY QUESTIONS

1 Copy and complete the table:

Anion	Test	Observations
a)	Add dilute acid	CO_2 gas produced
halide	Add nitric acid and silver nitrate solution	c) Chloride – d) Bromide – e) Iodide –
sulfate	b)	White precipitate of barium sulfate
nitrate	Add sodium hydroxide solution and aluminium powder, then warm gently	f)

2 Draw a flowchart to describe how to carry out a test for nitrate ions.

3 Compound A is a white solid which dissolves in water to produce a colourless solution. When this solution is acidified with nitric acid and silver nitrate is added, a white precipitate is produced. A flame test of A produces a bright red flame. Deduce the name of compound A and give your reasoning.

KEY POINTS

1 We identify carbonates by adding dilute acid, which produces carbon dioxide gas.
2 We identify halides by adding nitric acid and silver nitrate solution, which produce a precipitate of silver halide.
3 We identify sulfates by adding hydrochloric acid and barium chloride solution to produce a white precipitate of barium sulfate.
4 We identify nitrates by adding sodium hydroxide solution and a little aluminium powder to produce ammonia gas.

C3 5.3   Testing for organic substances

LEARNING OBJECTIVES

1 What simple test can we use to see if a substance is organic?
2 How can we detect carbon–carbon double bonds?
3 How can we find out the empirical formula of an organic compound by burning it?   [Higher]

From the earliest days of science people have heated substances to see what happened to them. Two hundred years ago a Swedish chemist called Jöns Jakob Berzelius decided that chemicals could be classified into two groups according to how they behaved when he heated them.

Chemicals that burned or charred on heating came mainly from living things, so Berzelius called these **organic** substances.

In contrast, other substances melted or vaporised when Berzelius heated them. These returned to their original state when they were cooled. Berzelius called these **inorganic** substances.

Our modern definition describes organic chemicals as substances which are based on the element carbon.

a) What is an organic substance?
b) How do organic substances behave when they are heated?

Detecting carbon–carbon double bonds (C=C)

As we saw on page 71, unsaturated hydrocarbons will react with bromine water, producing colourless compounds. This is a good way of testing a hydrocarbon to see if it contains a carbon–carbon double bond:

unsaturated hydrocarbon + bromine water → products
(colourless)   (orange-yellow) (colourless)

saturated hydrocarbon + bromine water → **no reaction**
(colourless)   (orange-yellow) (orange-brown)

c) How can we test for a carbon–carbon double bond?

This test is the basis for determining the number of carbon–carbon double bonds in unsaturated oils and fats. The oil is titrated against an iodine solution. (Iodine reacts with carbon–carbon double bonds in exactly the same way as bromine.) The 'iodine number' of the fat is based on the number of molecules of iodine needed to react with all of the carbon–carbon double bonds in one molecule of the fat.

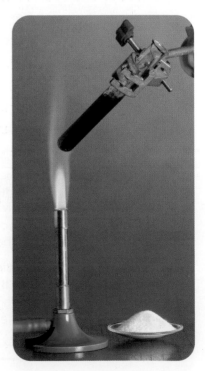

Figure 1 As a rule, organic substances char or burn when heated in air, while inorganic substances do not

HIGHER

Combustion analysis

We can find the empirical formula of an organic compound by burning it and measuring the amounts of the products formed.

Worked example

An organic substance Z contains carbon and hydrogen. A sample of Z is burnt in an excess of oxygen, producing 1.80 g of water and 3.52 g of carbon dioxide. What is the empirical formula of Z?

Solution

Step 1: Calculate moles of CO_2:

$$\text{The } M_r \text{ of } CO_2 \text{ is } 12 + (2 \times 16) = 44\,g$$

$$\text{Amount of } CO_2 = \frac{3.52}{44} = 0.08\,\text{moles}$$

Step 2: Calculate moles of H_2O:

$$\text{The } A_r \text{ of } H_2O \text{ is } (2 \times 1) + 16 = 18\,g$$

$$\text{Amount of } H_2O = \frac{1.80}{18} = 0.10\,\text{moles}$$

Each molecule of carbon dioxide formed requires one carbon atom from a molecule of Z. So for each mole of carbon dioxide formed, Z must contain one mole of carbon atoms.

Amount of C atoms in sample of Z = 0.08 mols

In the same way, each water molecule formed requires two hydrogen atoms from a molecule of Z. So for each mole of water formed, Z must contain two moles of hydrogen atoms.

Amount of H atoms in sample of Z = $0.10 \times 2 = 0.20$ mols

So Z contains carbon atoms and hydrogen atoms in the ratio $0.08 : 0.20 = 2 : 5$.

Therefore the empirical formula of Z is C_2H_5.

SUMMARY QUESTIONS

1 Copy and complete using the words below:

   burn   bromine   char   double

   Organic substances ...... or ...... when heated. Carbon–carbon ...... bonds (C=C) can be detected using ...... water.

2 Draw a flowchart for using combustion analysis to work out the empirical formula of a hydrocarbon. [Higher]

3 A hydrocarbon Y is burnt in excess oxygen. It produces 5.28 g of carbon dioxide and 2.16 g of water. What is the empirical formula of Y? [Higher]

4 Substance W contains carbon, hydrogen and oxygen. When 0.23 g of W is burnt in excess oxygen, 0.44 g of carbon dioxide and 0.27 g of water are produced. What is the empirical formula of W? [Higher]

KEY POINTS

1 Organic compounds burn or char when we heat them.
2 We can work out the empirical formula of an organic compound from the ratio of the products produced when it burns. [Higher]
3 We can detect carbon–carbon double bonds using bromine water.

C3 5.4 Instrumental analysis 1

Figure 1 Compared to methods 50 years ago, modern methods of analysis are quick, accurate and sensitive – three big advantages. They also need far fewer people, making them cheaper.

Many instrumental methods of detecting and identifying elements and compounds are in great demand in a number of different industries. Many industries require rapid and accurate methods for the analysis of their products. This can ensure that they are supplying a pure compound without any contamination by reactants or other by-products.

Similarly, society has become increasingly aware of the risk of pollution of both air and water. Careful monitoring of the environment using automated instrumental techniques has become more and more important.

Analysis and identification of different compounds and elements is also very important in health care. For example, in kidney dialysis machines aluminium can build up to dangerous levels. So the water used in these machines must be tested for aluminium at low concentrations. The presence of alcohol or other drugs in the blood stream can affect the way we react to other medical treatments. So detecting their presence in the blood is also very important.

The development of modern instrumental methods has been aided by the rapid progress made in technologies such as electronics and computing. These have enabled us to develop machines which can carry out the analysis and compare and make sense of the results. As these new technologies have emerged, new methods of analysing substances have been developed. These have a number of benefits over older methods:

● they are highly accurate,
● they are quicker, and
● they enable very small quantities of chemicals to be analysed.

Against this, the main disadvantages of using instrumental methods is that the equipment:

● is usually very expensive,
● takes special training to use it, and
● gives results that can often be interpreted only by comparison with already available known specimens.

a) What has aided the development of instrumental methods of chemical analysis?
b) What benefits do these methods have over older methods of analysis?

HIGHER

Detecting elements

There are two principal methods that are used for detecting and identifying elements. ***Atomic absorption spectroscopy (AAS)*** is a technique used to measure the concentration of a particular metal in a liquid sample.

The liquid sample is fed into the flame where it vaporises. Light passing through the flame passes through a monochromator which selects the wavelength to be studied. This light then falls on a detector which produces an electric current that depends on the light intensity. The electrical current is amplified and processed by the instrument's electronic circuits. This provides a measurement of the concentration of the metal in the sample.

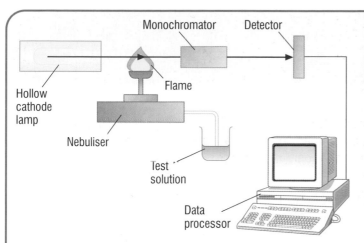

Figure 2 Atomic Absorption Spectroscopy (AAS) is a powerful technique for measuring the amount of particular metals in a sample

c) What is atomic absorption spectroscopy used for?

The other method uses an instrument called a **mass spectrometer** to compare the mass of different atoms. This provides an important way of determining relative atomic masses, as well as identifying the particular elements present in a sample.

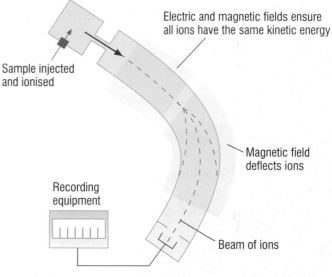

Figure 3 A mass spectrometer provides an accurate way of measuring the mass of atoms

SCIENCE @ WORK

Atomic absorption spectroscopy is used in the steel industry to measure the levels of other metals present in steel. It is also used in medicine, to measure the levels of metal ions in a person's blood.

N.B. You don't need to remember the details of how atomic absorption or mass spectrometers work.

SUMMARY QUESTIONS

1 Copy and complete using the words below:

   analysis   medicine   electronics   industry

   Modern instrumental methods of chemical .... depend on .... They are used widely in areas including .... and ....

2 What are the main advantages and disadvantages of using instrumental analysis compared to traditional practical methods?

3 How are advances in technology linked to developments in methods of instrumental analysis?

4 Name two methods used to identify elements.   [Higher]

KEY POINTS

1 Modern instrumental techniques depend on electronics and computers to provide fast, accurate and sensitive ways of analysing chemical substances.

2 Atomic absorption spectroscopy (AAS) and mass spectrometry can be used to analyse and identify the chemical elements in a sample.   [Higher]

C3 5.5

Instrumental analysis 2

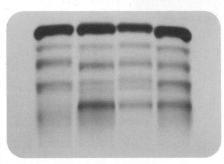

1 How can we identify compounds?

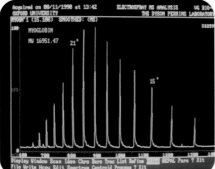

Figure 1 Gel electrophoresis is another type of chromatography. It uses an electric field to separate compounds moving across a gel-covered plate

Mass spectrometry can be used to provide information about compounds as well as elements. The mass spectrum of a molecule tells us its relative mass, and can also provide information about its structure because some of the molecules break up as they go through the spectrometer.

There is a very wide range of chemical instruments that we can use to identify unknown chemical compounds. In many cases the instruments use techniques that are simply more sophisticated and automated versions of techniques that we use in the school chemistry lab.

One example of this is **chromatography**, which is used to separate different compounds within a mixture. The technique is based on how well they dissolve in a particular solvent. This determines how far each substance travels across a stationary medium such as a piece of chromatography paper.

We can use:

- *gas-liquid chromatography* to separate compounds that are easily vaporised,
- *gel permeation chromatography* to separate compounds according to the size of their molecules,
- *ion-exchange chromatography* to separate compounds containing differently charged particles, and
- *high-performance liquid chromatography* to separate compounds in solution.

Once the compounds in a mixture have been separated, they may be identified by comparison with the distance moved by known substances.

Alternatively, other instruments may then be used to identify them.

Figure 2 Athletes train for years to reach the peak of their performance in the Olympic Games. Instruments are used to carry out rapid automated analysis of the test samples collected during the competition, detecting the presence of illegal drugs that could give an athlete an unfair advantage.

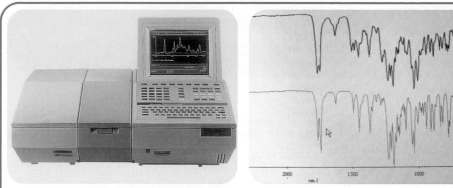

The frequency of infra-red radiation is lower than that of visible light. Molecules can absorb the energy carried by infra-red radiation, which makes the bonds in the molecule vibrate. The frequency of the infra-red radiation that a molecule absorbs depends on the chemical bonds in it – so the **infra-red spectrum** of a compound can provide valuable information about its structure.

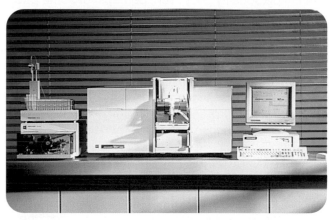

The energy carried by visible and ultraviolet light may also be absorbed by compounds, promoting electrons to higher energy levels. **UV–visible spectroscopy** is used with molecules and inorganic ions. It provides us with very limited information about the structure of compounds but is very useful when we want to make quantitative measurements about the amount of a substance that is present, based on the amount of light that a sample absorbs at a particular frequency.

The nuclei of some atoms behave like tiny magnets, and can be affected by a magnetic field. **NMR (nuclear magnetic resonance) spectroscopy** uses radio waves to 'flip' these nuclei in a molecule between different alignments within a strong magnetic field, rather like changing the alignment of a compass needle by pushing it with your finger. If we measure the amount of energy needed to cause this flipping at different frequencies of radio waves, the spectrum produced tells us a great deal about the way that the atoms in the molecule are arranged. NMR spectroscopy is particularly useful for determining the structure of large organic molecules. It is also the basis of the medical imaging technique which we call magnetic resonance imaging.

SUMMARY QUESTIONS

1 Copy and complete using the words below:

**chromatography   energy   formula   mass spectrometry
mixture**

Separating a …… of compounds uses techniques which are based on ……. . Identifying compounds once they have been separated then uses techniques like …… …… , in which the relative …… mass of the molecule is measured, or other techniques in which the …… absorbed by the compound is measured.

2 Which of the techniques described on these two pages is **not** used to determine the structure of molecules? Explain your answer.

KEY POINTS

1 Compounds in a mixture can be separated using chromatography.
2 Once separated, compounds can be identified using a variety of instrumental techniques.

C3 5.6 | Chemical analysis

Paper, ink and forgery

In Germany in 1983 *Stern* magazine published extracts from what they claimed were the diaries of Adolf Hitler. The magazine had paid about $6 million for sixty small handwritten books, which covered the period from 1932 to 1945. At a press-conference in April 1983 the historian Hugh Trevor-Roper said: 'I am now satisfied that the documents are authentic; that the history of their wanderings since 1945 is true; and that the standard accounts of Hitler's writing habits, of his personality and, even, perhaps, of some public events, may in consequence have to be revised.'

But many people were suspicious, and within two weeks the police were on the case. Using chemical analysis, forensic scientists were able to:

- show that the paper on which the diaries were printed contained *blankophor*, a fluorescent whitening substance that was not used in paper until after 1954;
- prove that polyester and viscose, modern polymers, were present in the bindings of the books;
- use chromatography to prove that none of the four different inks used in the diaries were available during Hitler's lifetime.

The diaries were actually written by Konrad Kujau, a well-known forger. Both Kujau and Gerd Heidemann – the journalist who 'discovered' the diaries – were sent to prison for $3\frac{1}{2}$ years.

ACTIVITY

Using the information from this chapter, and any further information you can find, write a newspaper article entitled 'Chemical analysis in action – can we believe our eyes?'

Doping – drugs in sport

The doping classes and methods that are prohibited by most governing bodies of sport are based on those of the World Anti-Doping Agency. However, it is important to note that some of the prohibited substances may vary from sport to sport. It is the athlete's responsibility to know their sport's anti-doping regulations. In cases of uncertainty, it is important to check with the appropriate governing body or UK Sport. Also be sure to read carefully the anti-doping rules adopted by the relevant governing body and international sports federations.

Athletes are advised to check all medications and substances with their doctor or governing body medical officer. All substances should be checked carefully when travelling abroad as many products can, and do, contain different substances to those found in the UK.

ACTIVITY

Chemists are active in the fight to uncover the abuse of drugs in sport. Design a poster which could be displayed in the changing rooms of a sports centre used by young athletes to show them how chemical analysis using sensitive instruments is used to catch the 'drug cheats'.

Teen*zine*

A Difficult Choice?
the BIG debate

It can be a tough road to the top in any field – and nowhere is it tougher than in athletics. The route from the school athletics field to the Olympic gold medal winners' podium is full of hard work, and you may need to make some difficult choices too.

So how far would you go to get into a sports team?

- try harder in school sports lessons
- go along to training sessions outside school
- trip up competing athletes to try to injure them
- get your parents to pay for private coaching
- skip homework to work in a fast-food restaurant to earn money to pay for private coaching
- steal money to pay for private coaching
- follow a special diet
- take a 'herbal supplement' that your coach recommends to improve your performance
- take drugs which a fellow competitor says doubled her performance

When we are faced with decisions like these we need to think about what we *can* do. Once we know this, we must decide what we *should* do.

Science is very useful for the first type of question. For example, if we want to know whether drugs or diet can affect our performance as an athlete, science can provide answers. Science can also tell us whether a particular substance is likely to be bad for us.

But science is not helpful when we want to know how we *should* behave – so we need to look elsewhere. One way of helping us to make decisions about our behaviour is to think about *goals*, *rights* and *duties*.

A *goal* is something we want to achieve – for example, to be the fastest person in the 200 metres. One way of judging someone's behaviour is to look at why they behave in the way they do – in other words, to look at their goals.

Our *rights* refer to the way that we can expect to be treated – for example, we may be able to say what we think, a right that is called our *freedom of speech*.

And *duties* describe the way that we should behave – for example, the duty to tell the truth.

Duties, rights and goals are often linked. If you are accused of a crime you have a *right* to a lawyer who has a *duty* to try to get you acquitted – which is the lawyer's *goal*.

But it can get complicated. What about the case of a dying patient, who asks his doctor not to keep him alive any longer? Does the doctor have a *duty* to do as the patient asks because that patient has the *right* to decide how and when to die? Or is the doctor's *duty* to ignore the patient's wishes because of the *goal* of preserving life?

ACTIVITY

James is a member of the school swimming squad. Nelson wants to get into the squad. Draw up a table to show the *goals, rights* and *duties* that you think James and Nelson each have. How do you think that these two young people *should* behave. What advice would *you* give them and why?

SUMMARY QUESTIONS

1 Copy and complete the table.

Add dilute acid	Add sodium hydroxide solution and warm	Add sodium hydroxide solution and add aluminium – then warm	Flame test	Substance
Nothing observed	Nothing observed	Gas evolved turns damp red litmus paper blue after aluminium added	Golden yellow	e)
Fizzing – gas turns lime water milky	Gas evolved turns damp universal indicator paper blue	Nothing observed	Nothing observed	f)
a)	b)	c)	d)	Calcium carbonate

2 A technique called **gel electrophoresis** is used to analyse DNA. Analysing DNA using this technique produces a plate which carries a series of bands, according to the composition of the DNA:

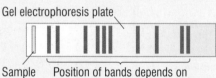

Gel electrophoresis plate

Sample added   Position of bands depends on composition of DNA in sample

Scientists analysed DNA from three suspects in a criminal investigation using this method and compared it with the DNA from a sample found at the scene of a crime. The following results were obtained:

DNA from crime scene

Suspect A

Suspect B

Suspect C

a) How is evidence like this used to decide whether someone accused of a crime is guilty or not?

b) According to the results, which of the suspects A, B or C could have been present at the scene of the crime? Explain your answer.

c) It might not be a good idea to rely on this kind of evidence alone when deciding if someone is guilty of a crime or not. Why?   [Higher]

EXAM-STYLE QUESTIONS

1 Match each compound in the list with the test and result in (a) to (d).

copper carbonate   magnesium sulfate
potassium nitrate   sodium chloride

(a) Add dilute hydrochloric acid and barium chloride solution. A white precipitate forms.

(b) Add dilute hydrochloric acid. Carbon dioxide gas is given off.

(c) Add dilute nitric acid and silver nitrate solution. A white precipitate forms.

(d) Add sodium hydroxide solution and aluminium powder. The gas given off turns litmus paper blue.   (4)

2 A technician investigated a mixture that had been left in the laboratory. The mixture contained a black solid, A, and some colourless crystals, B.

(a) The mixture was added to water, stirred and then filtered. Solid A was left in the filter paper and a solution of B was collected.

(b) Solid A was washed with water and added to warm nitric acid. A blue solution C was produced.

(c) Sodium hydroxide solution was added to C. A blue precipitate formed.

(d) Some of the solution of B was sprayed into a Bunsen burner flame and produced a lilac colour.

(e) Dilute hydrochloric acid and barium chloride were added to the solution of B. A white precipitate formed.

Write deductions for each of the steps (a) to (e). Identify A, B and C.   (8)

3 Match the instrumental methods V–Z in the table with the analyses described in (a) to (e).

Separating methods	V	Mass spectrometry	Sorts and identifies atoms, molecules and fragments of molecules.
	W	Gas–liquid chromatography	Separates and finds the amounts of molecules in a homologous series.
Spectral methods	X	Infra-red spectrometry	Identifies molecules and particular bonds within molecules.
	Y	Ultraviolet spectroscopy	Used to measure the concentration of molecules in solutions.
	Z	Atomic absorption spectroscopy	Determines the presence and concentration of metals in liquid samples.

(a) Finding the concentration of caffeine in a soft drink.

(b) Finding the formula of an organic compound.

(c) Finding the amounts of alkanes in petrol.

(d) Finding the amount of lead in a sample of water.

(e) Showing which trans fatty acids are in margarine.

(5)

[Higher]

4 Read the information about steel and answer the questions (a) to (c).

> Very small changes in the amounts of elements in steel can affect its properties. Atomic emission spectroscopy is an instrumental method of analysis that helps to ensure steels have exactly the correct composition. A small sample of the steel from the furnace is placed in the instrument. Each element present emits light of a particular wavelength and the intensity of the light is proportional to the concentration of the element in the steel. A computer processes the results so that the steel-maker knows almost instantly and very accurately the composition of the steel. Any adjustments can be made before the steel is poured out of the furnace.

(a) What does atomic emission spectroscopy detect and measure? (2)

(b) Give **three** reasons why atomic emission spectroscopy is more useful in the steel industry than traditional laboratory methods of analysis. (3)

(c) Suggest one advance in technology that has helped the rapid development of instrumental methods of analysis. (1)

[Higher]

HOW SCIENCE WORKS QUESTIONS

Atomic absorption spectroscopy and the space shuttle

One of the world's most amazing sights is the space shuttle taking off from Cape Canaveral Space Center. The rockets must deliver the shuttle into orbit. It is essential that they are reliable and that they are re-usable. They are therefore tested to establish their reliability. However, like a car engine they do show signs of wear after use. Atomic absorption spectroscopy (AAS) can identify that wear, by analysing the exhaust gases from the rockets.

AAS will identify the presence of iron, nickel, cobalt, copper and silver. Any increase in these elements suggests engine wear.

a) Who should carry out these tests? Explain your answer. (2)

b) How might the invention of AAS be used to improve engine design for the space shuttle? (1)

c) How else might this technology be used? (1)

d) On 28 January 1986 the Challenger space shuttle blew up just over a minute into its flight. An 'O ring' on the solid rocket booster had broken. All seven crew members were killed.

AAS can only detect engine wear. It is for the judgement of the scientists to decide if that wear is dangerous. Make the decision too soon and space travel becomes too expensive. Make the decision too late and astronauts die. What are the ethical issues around such a decision? (2)

EXAMINATION-STYLE QUESTIONS

1   In the modern periodic table, the elements in periods 2 and 3 have similar chemical properties every ninth element.

See pages 210–13, 218

(a)   Complete the sentence: In the modern periodic table, elements are arranged in order of their ......   *(1 mark)*

(b)   Complete the sentence: The chemical properties of an element depend upon the number of ...... in the highest energy level (outer shell) of its atoms.   *(1 mark)*

(c)   Explain why the chemical properties are repeated every ninth element in these two periods.   *(2 marks)*

(d)   Why did Newlands and Mendeleev find that the properties repeated every eighth element?   *(1 mark)*

2   $25.0\,cm^3$ of sodium hydroxide solution reacted with $16.0\,cm^3$ of $0.50\,mol$ per dm^3 sulfuric acid. The equation for the reaction is:

See pages 226–9

$$2NaOH + H_2SO_4 \rightarrow Na_2SO_4 + H_2O$$

(a)   Name the method that could be used to find these reacting volumes.   *(1 mark)*

(b)   Name the apparatus you would use to add the acid to the alkali.   *(1 mark)*

(c)   What would you need to add to the solution to find the end point of the reaction?   *(1 mark)*

(d)   Explain why it would be advisable to repeat the method more than once.   *(1 mark)*

(e)   Calculate the number of moles of sulfuric acid in $16.0\,cm^3$ of the solution.   *(2 marks)*

(f)   How many moles of sodium hydroxide were in $25.0\,cm^3$ of solution?   *(1 mark)*

(g)   Calculate the concentration of the sodium hydroxide solution in moles per dm^3.   *(2 marks)*

(h)   Calculate the concentration of the sodium hydroxide solution in grams per dm^3.   *(2 marks)*
      (A_r values: Na = 23, O = 16, H = 1)   [Higher]

3   The table gives information about some fuels. All of the fuels, except for hydrogen, are organic compounds.

See pages 248–9

Name of fuel	Energy produced per gram (kJ/g)	Volume of 1 gram at 20°C and 1 atm (cm³)
Coal	25	1.35
Ethanol	27	0.79
Hydrogen	120	12 000
Petrol	44	0.72
Sugar	17	1.58
Vegetable oil	38	0.93
Wood	15	0.70

(a)   Which fuel produces the most energy per gram?   (1 mark)

(b)   Which of the liquid fuels produces most energy per cm^3?   (1 mark)

(c)   Which of the solid fuels produces the least energy per cm^3?   (1 mark)

(d)   Suggest **two** reasons why liquid fuels are used for road vehicles.   (2 marks)

(e)   Hydrogen could be used as a fuel for cars.

   (i)   Suggest **two** reasons why hydrogen would be a good choice of fuel.
                                                              (2 marks)

   (ii)   Suggest **two** reasons why hydrogen is not more widely used as a fuel for cars.   (2 marks)

(f)   Describe an experiment you could do to compare the energy produced by the liquid fuels in the table. Give only the essential steps and explain how you would make the experiment a fair test. Explain why it is not possible to obtain accurate values for the energy produced by the fuels in your experiment.   (6 marks)

4   An organic compound contains carbon and hydrogen only. A sample of the compound was burnt completely. It produced 1.32 g of carbon dioxide and 0.72 g of water. What is the empirical formula of the compound? Show all of your working.   (4 marks)
   (A_r values: C = 12, O = 16, H = 1)   [Higher]

See page 263

Glossary

A

Accuracy An accurate measurement is close to the true value.

Acid A substance that produces hydrogen ions when it dissolves in water.

Acid rain Rain that is acidic due to dissolved gases, such as sulphur dioxide, produced by the burning of fossil fuels.

Activation energy The minimum amount of energy needed for a given chemical reaction to take place.

Activity Number of atoms of a radioactive substance that decay each second.

Alkali A soluble base.

Alkali metals A group of soft metals with one electron in the outer shell, which react with water to form alkaline solutions.

Alkane A hydrocarbon with the general formula C_nH_{2n+2}.

Alkene A hydrocarbon with the general formula C_nH_{2n}.

Alloy A metallic substance formed by combining two or more metals.

Anhydrous An anhydrous substance does not contain water.

Anode Positive electrode.

Anomalous A measurement that is well away from the pattern shown by other results.

Aqueous solution A solution with water as the solvent.

Association When two variables change together, but they are both linked by a third variable. E.g. lack of carbon dioxide in soil and poor growth of plants: both could be linked to too much water in the soil.

Atom The smallest part of an element.

Atom economy The efficiency of a chemical reaction in terms of all of the atoms involved.

Atomic nucleus Positively charged object composed of protons and neutrons at the centre of every atom with one or more electrons moving round it.

Atomic number The number of protons in a nucleus, symbol Z (also called the proton number).

B

Bar charts Used when the independent variable is categoric and the dependent variable is continuous.

Bases Compounds which react with acids to neutralise them.

Bias The influence placed on scientific evidence because of: wanting to prove your own ideas; supporting the person who is paying you; political influence; the status of the experimenter.

Biodiesel Diesel fuel made from plant materials.

Blast furnace A reaction vessel in which iron oxide is heated with coke and limestone to produce iron.

Bomb calorimeter Apparatus used to measure the enthalpy change of a chemical reaction.

Bond energy The energy needed to break a particular chemical bond.

Bronze An alloy containing copper and tin.

Burette A glass tube with markings and a tap used to add precisely known amounts of liquids to a container.

C

Cast iron Iron containing between 2% and 5% carbon.

Catalyst A substance that speeds up the rate of another reaction but is not used up or changed itself.

Categoric variable These tell us the name of the variable, e.g. copper, iron, magnesium.

Cathode Negative electrode.

Causal link One change in a variable has caused a change in another variable. You can only be reasonably certain of this when you have valid and reliable evidence. E.g. increasing temperature causes an increase in the rate of the reaction.

Cement A building material made from limestone and clay mixed with water.

Chain reaction Reactions in which one reaction causes further reactions, which in turn cause further reactions, etc. A nuclear chain reaction occurs when fission neutrons cause further fission, so more fission neutrons are released. These go on to produce further fission.

Chance When there is no scientific link between the two variables. E.g. increased sea temperatures and increased diabetes.

Chloride ion A chlorine atom that has gained one electron, which gives it a negative charge.

Chromatography A technique used to separate a mixture of substances using a stationary and a moving phase.

Collision theory An explanation of chemical reactions in terms of reacting particles colliding with sufficient energy for a reaction to take place.

Combustion The process of burning.

Compound A substance made of two or more types of atom chemically joined together.

Concentration gradient The gradient between an area where a substance is at a high concentration and an area where it is at a low concentration.

Conclusion A conclusion considers the results and states how those results match the hypothesis. The conclusion must not go beyond the data available.

Concrete A building material made from sand, cement and crushed rock mixed with water.

Conduction Heat transfer in a substance due to motion of particles in the substance.

Conduction electrons Electrons that move about freely inside a metal because they are not attached to individual atoms.

Conservation of energy Energy cannot be created or destroyed.

Continuous variable A continuous variable can be any numerical value, e.g. temperature.

Control groups Often used when there are a large number of control variables that cannot be kept constant. E.g. when testing a drug on thousands of different people, half will be given the drug and half will be given a similar treatment that does not contain the drug (placebo).

Control variable These are the variables that might affect your result and therefore must be kept the same for a valid investigation. E.g. volume of acid used in an investigation of the effect of temperature.

Controlled An experiment is controlled when all variables that might affect your result (apart from the independent variable) have been kept constant.

Convection Heat transfer in a liquid or gas due to circulation currents.

Convection currents The flow of a fluid due to differences in temperature. E.g. circulation of the upper part of the Earth's mantle.

Core The central part of the Earth below the mantle.

Covalent bonds The bonds formed when atoms join together by sharing electrons.

Cracking Breaking a molecule apart using heat.

Crust The outermost layer of the Earth.

D

Data Measurements or observations of a variable. Plural of datum.

Decompose To split up.

Delocalised electrons Electrons in a molecule which do not belong to a single atom or a single bond.

Density Mass per unit volume of a substance.

Dependent variable The variable that you are measuring as a result of changing the independent variable, e.g. the volume of CO_2 produced.

Diffusion The net movement of particles of a gas or a solute from an area of high concentration to an area of low concentration (along a concentration gradient).

Directly proportional A graph will show this if the line of best fit is a straight line through the origin.

Discrete variable These are numerical, but can only be whole numbers, e.g. numbers of layers of insulation.

Dissociation The separation of a substance into two or more simpler substances, or of a molecule into atoms or ions.

E

E number A number given to a food additive in order to identify it.

Economic How science affects the cost of goods and services. E.g. developing the use of catalysts might decrease the cost of production.

Elastic A material is elastic if it is able to regain its shape after it has been squashed or stretched.

Electrolyte A substance that conducts electricity when molten or when dissolved in water.

Electron microscope An instrument used to magnify specimens using a beam of electrons.

Electronic structure The arrangement of electrons around the nucleus of an atom.

Electrons Negative particles found outside the nucleus of an atom.

Element A substance made up of only one type of atom.

Empirical formula A chemical formula that shows the ratio of the number of atoms in a compound.

Emulsifier A substance which stops the two liquids in an emulsion separating.

Emulsion A mixture of tiny droplets of one liquid in another liquid.

End point The point in a titration where the reaction is complete and titration should stop.

Endothermic Involving a net absorption of energy.

Energy level See **shells**.

Energy transfer Energy transferred from one place to another.

Ethical Whether it is 'right' or 'wrong' to do something. E.g. experimentation on animals to develop new drugs.

Equilibrium The point at which a reversible reaction takes place at exactly the same rate in both directions.

Evidence Scientific evidence should be reliable and valid. It can take many forms. It could be an observation, a measurement or data that somebody else has obtained.

Exothermic Involving a net release of energy.

F

Fair test Only the independent variable is affecting your dependent variable, all other variables are kept the same.

Fatty acids Building blocks of lipids.

Fluid A liquid or a gas.

Food additive A substance added to food to improve its flavour, texture or shelf-life.

Fossil fuel Coal, oil or gas or any other fuel formed long ago from the fossilised remains of dead plants or creatures.

Fractional distillation A way of separating a mixture of substances according to their different boiling points.

Free electrons Electrons that move about freely inside a metal and are not held inside an atom.

Fullerene A type of giant structure made up of carbon atoms.

G

Galvanising Covering iron with a protective layer of zinc.

Gasohol A mixture of petrol (gasoline) and ethanol.

Giant covalent structures Giant structures held together by many covalent bonds which give them high melting points and hardness, e.g. diamond.

Giant structure Large numbers of atoms or ions arranged in a regular way.

Glass Transparent material made by heating a mixture of sand, sodium carbonate and limestone.

Global dimming A gradual reduction in the amount of light reaching the Earth's surface.

Global warming Warming of the Earth due to greenhouse gases in the atmosphere trapping infra-red radiation from the surface.

Glycerol Building block of lipids.

Grain A metal crystal.

Grain boundaries Where two metal crystals meet.

Greenhouse gases Gases such as carbon dioxide in the atmosphere that absorb infra-red radiation from the Earth's surface.

Group A vertical column of elements in the periodic table.

H

Haber process The industrial process used to make ammonia.

Haematite An ore containing iron combined with oxygen.

Half-life of a radioactive isotope Time taken for the number of nuclei of the isotope (or mass of the isotope) in a sample to halve.

Hardening Adding hydrogen to an oil, replacing the carbon–carbon double bonds in the molecules of an oil with carbon–carbon single bonds.

Hard water Water containing dissolved calcium and/or magnesium salts.

Hydrated A hydrated substance contains water.

Hydrocarbon A compound containing only carbon and hydrogen.

Hypothesis Using theory to suggest explanations for observations, e.g. 'I think that the change in colour is caused by copper ions.'

I

Independent variable The variable that you have decided to change in an investigation, e.g. temperature of the acid.

Indicator A chemical compound that changes colour according to the pH of the solution it is in.

Inert Unreactive.

Inorganic Substances that consist principally of elements other than carbon.

Insoluble Unable to dissolve in a given solvent.

Intermolecular forces Forces of attraction between molecules.

Interval measurements The values of your independent variable that you choose within the range e.g. $10\,cm^3$; $20\,cm^3$; $30\,cm^3$; $40\,cm^3$; $50\,cm^3$.

Ionic bond A chemical bond formed when one atom gives up one or more electrons to another atom.

Ionisation Any process in which atoms become charged.

Ionising radiation Radiation that ionises substances it passes through. Alpha, beta, gamma and X-radiation are all ionising.

Ion A charged atom.

Isotopes Atoms of the same element which have different numbers of neutrons in their nuclei.

L

Law of force between charged objects Like charges repel; unlike charges attract.

Lime water Solution of calcium hydroxide, used to test for carbon dioxide.

Limiting factors Factors which limit the rate of a reaction, e.g. photosynthesis.

Line graphs Used when the independent and the dependent variables are both continuous.

Line of best fit Used to show the underlying relationship between the independent and the dependent variables. It should fit the pattern in the results and have roughly the same number of plots on each side of the line. It could be a straight line or a curve. Remember to ignore any anomalies!

Linear These are straight line graphs that can be positive (as the concentration increases so too does the oxygen produced) or negative (as the concentration increases the oxygen produced decreases).

Link due to association When two variables change together, but they are both linked by a third variable. E.g. less oxygen dissolved in water and more sodium chloride dissolved, both could be due to the higher temperature of the water.

Link due to chance When there is no scientific link between the two variables. E.g. increased sea temperatures and increased diabetes.

Lipids Fats and oils.

M

Magnesium A metallic element. Magnesium ions are needed by plants to make chlorophyll.

Malleable Capable of being hammered into shapes without smashing – a property of metals.

Mantle The layer of the Earth between the crust and the core.

Mass number The total number of protons and neutrons in the nucleus of an atom (symbol A).

Mass spectrometer An instrument used to measure the mass of atoms and molecules.

Mean Add up all of the measurements and divide by how many measurements there are. Don't forget to ignore any anomalous results.

Mixing occurs when two or more substances are physically mixed but not chemically combined.

Model Description of a theory or theories that suggests further ideas that could test those theories. E.g. 'plum pudding' model of the atom that was tested and found not to be correct. A better model was then suggested.

Mole The relative formula mass of a substance in grams.

Molecular formula A formula that shows the total number of the different kinds of atoms in a molecule.

Monomer A molecule that can combine with other, similar, molecules to form a polymer.

Mortar Mixture of sand, cement and water used to hold building materials together.

N

Net Overall.

Neutral Neither acid nor alkaline.

Neutrons Neutral particles found in the nucleus of an atom.

Nitrogen Inert gas making up around 80% of the Earth's atmosphere.

Noble gas One of the six unreactive gases found in group 0 of the periodic table. They have a complete outer shell of electrons, e.g. neon, argon, helium.

Nuclear model of the atom Every atom contains a positively charged nucleus consisting of neutrons and protons. This is where most of its mass is concentrated, and it is much smaller than the atom. Electrons move about in the space surrounding the nucleus.

O

OILRIG Oxidation Is Loss, Reduction Is Gain (of electrons).

Opinion Opinions are personal judgements. Opinions can be formed from scientific evidence or non-scientific ideas.

Ordered variable Variables that can be put into an order, e.g. small, large, huge lumps of rock.

Ores Rocks that contain enough metal to make it economical to extract the metal.

Organic substances contain (mainly) carbon in combination with other elements.

Osmosis The net movement of water from an area of high concentration (of water) to an area of low concentration (of water) along a concentration gradient.

Oxidation Losing electrons.

Oxidised See **oxidation**.

Ozone layer Layer of ozone gas in the Earth's atmosphere that absorbs ultraviolet radiation.

P

Partially permeable Allowing only certain substances to pass through.

Pay-back period (or time) Length of time for the savings from an improvement to match the actual cost of the improvment.

Percentage yield The percentage of product formed in a chemical reaction compared with the maximum possible amount of product that could be formed.

Period A horizontal row of elements in the periodic table.

pH scale A scale running from 0 to 14 that describes the degree of acidity of a solution.

Pipette A glass tube used to measure precise volumes of liquids.

Plum pudding model of the atom A model of the atom which supposed that the positive charge was evenly spread throughout its matter and the negative charge was held in tiny particles (electrons) inside the atom.

Plasma A gas consisting of bare nuclei (i.e. atoms stripped of their electrons).

Pollution The contamination of air, water or soil by substances which are harmful to living organisms.

Polymer A substance consisting of very large molecules made of smaller identical molecules called monomers.

Precipitate A solid material produced from a solution.

Precipitation See **precipitate**.

Precision Where your repeat readings are very close to each other. This is related to the smallest scale division on the measuring instrument used.

Prediction A hypothesis that can be used to design an investigation e.g. I predict that if I increase the concentration of hydrochloric acid, there will be an increase in the volume of carbon dioxide produced.

Proton acceptor A modern definition of an alkali.

Proton donor A modern definition of an acid.

Proton number See **atomic number**.

Protons Positive particles found in the nucleus of an atom.

Q

Quicklime Calcium oxide.

R

Random changes Changes that cannot be predicted.

Random error Measurements when repeated are rarely exactly the same. If they differ randomly then it is probably due to human error when carrying out the investigation.

Range The maximum and minimum values.

Redox A **RED**uction **OX**idation reaction in which electrons are lost by one substance and gained by another.

Reduced See **reduction**.

Reduction Gaining electrons.

Reduction reaction A reaction in which an atom or ion gains electrons.

Relative atomic mass The mass of an atom compared with an atom of $^{12}_{6}C$. This is usually the same as or very similar to the mass number of the element.

Relative formula mass The mass of a chemical compound based on the relative atomic masses of the elements involved.

Reliable Describes data we can trust. E.g. others can get the same results, even using different methods.

Reliability The trustworthiness of data collected.

Renewable energy Energy from sources that never run out, including wind energy, wave energy, tidal energy, hydroelectricity, solar energy and geothermal energy.

Reversible reaction A reaction in which the products immediately react together to produce the original reactants.

S

Saturated A hydrocarbon which contains as many hydrogen atoms as possible in each molecule.

Saturated solution A solution in which no more solute will dissolve.

Scale The substance formed when hard water is boiled.

Scattergrams Used when you want to see how variables relate to each other. E.g. the depth that the ore has been mined and the yield of the metal.

Scum The substance formed when soap reacts with hard water.

Sensitivity The smallest change that an instrument can measure, e.g. 0.1 mm.

Shells The region in which electrons are concentrated as they travel around the nucleus of an atom.

Slag The waste produced when iron is made in a blast furnace.

Slaked lime Calcium hydroxide.

Smart alloy An alloy which returns to its original shape when it is heated.

Social issues How science influences and is influenced by its effects on our friends and neighbours. E.g. building an oil storage depot next to a village.

Stabiliser A substance with molecules that produce large 'cages' full of air when they are mixed with water.

Steels Alloys of iron containing controlled amounts of carbon and/or other metals.

Sodium ion A sodium atom that has lost an electron to give it a positive charge.

Soft water Water containing no dissolved calcium and/or magnesium salts.

Soluble Able to dissolve in a given solvent.

Solubility The extent to which one substance will dissolve in another.

Solubility curve A graph describing the solubility of a substance.

Solute The solid which dissolves in a solvent to form a solution.

Solvent A liquid in which another substance can be dissolved to make a solution.

Strong acid/alkali Acid/alkali that is (almost) completely dissociated when dissolved in water.

Sugars Simple carbohydrates.

Systematic error If the data is inaccurate in a constant way, e.g. all results are 10 mm more than they should be. This is often due to the method being routinely wrong.

T

Tectonic plates Huge sections of the Earth's crust and upper mantle.

Technology Scientific knowledge can be used to develop equipment and processes that can in turn be used for scientific work.

Theory A theory is not a guess or a fact. It is the best way to explain why something is happening. E.g. Sea levels are rising, and the global warming theory is the best way to describe why they are. Theories can be changed when better evidence is available.

Thermal decomposition Splitting up a substance by means of heat.

Thermosetting A polymer that hardens or sets permanently when it is formed by heating the monomers of which it is made.

Thermosoftening A polymer that softens when it is heated.

Titration A method for measuring the amount of substance in a solution.

Transition elements See *transition metals*.

Transition metals The large block of metallic elements in the middle of the periodic table.

Trial run Carried out before you start your full investigation to find out the range and the interval measurements for your independent variable.

Tsunami A large wave caused by an underwater earthquake or volcanic eruption.

U

Universal indicator A substance containing a range of indicators to provide a measurement of pH.

Unsaturated A hydrocarbon which contains a carbon–carbon double bond.

Unsaturated oils Oils in which the molecules contain carbon atoms joined together by carbon–carbon double bonds (C=C).

V

Valid Describes an investigation that successfully gathers the data needed to answer the original question. Data may not be valid if you have not carried out a fair test.

Valid data Evidence that can be reproduced by others and answers the original question. Data may not be valid if you have not carried out a fair test.

W

Water cycle The continuous process by which water is distributed throughout the Earth and its atmosphere.

Weak acid/alkali Acid/alkali that is only slightly dissociated when dissolved in water.

Y

Yield The amount of product formed in a chemical reaction.

Z

Zero error A systematic error, often due to the measuring instrument having an incorrect zero. E.g. forgetting that the end of the ruler is not at zero.

Index

Acknowledgements

Alamy/Adrian Sherratt 31.3, /Education Photos 164.1, /Holt Studios International Ltd 200.2; ALCAN 136.2b; Arm & Hammer 241.2; B Drake/Photodisc 19 (NT) 205r; Ben Keiser 151.1; Bibliotheque de L'Ecole des mines de Paris 211.4a, 211.4b; Blueflag 203m; Brita 242.2; Corbis V257 (NT) 74.2; Corbis/Jose Luis Pelaez 76.2; Corel 41 (NT) 26.4, 254bl; Corel 62 (NT) 248.1b; Corel 149 (NT) 174.2; Corel 220 (NT) 255m; Corel 245 (NT) 125.3; Corel 250 (NT) 87.3, 209br; Corel 284 (NT) 3.1, 162.1; Corel 318 (NT) 44.1; Corel 320 (NT) 38.2; Corel 340 (NT) 123.6; Corel 357 (NT) 242.1; Corel 366 (NT) 18.2; Corel 437 (NT) 26.3, 33.4; Corel 448 (NT) 5.2; Corel 454 (NT) 255tr; Corel 456 (NT) 103.4; Corel 467 (NT) 168.1; Corel 501 (NT) 112tl; Corel 587 (NT) 26.1; Corel 589 (NT) 240.1; Corel 603 (NT) 255mr; Corel 625 (NT) 149.2; Corel 632 (NT) 41.3; Corel 674 (NT) 235.3; Corel 696 (NT) 254tr; Corel 710 (NT) 174.1; Corel 771 (NT) 266.2; CSIRO 47.4; David Buffington/Photodisc 46 (NT) 238.2; Digital Vision 1 (NT) 40.1, 42.1, 178.1; Digital Vision 6 (NT) 271tr; Digital Vision 7 (NT) 97.2, 234.1; Digital Vision 14 (NT) 68tl, 100.1; Digital Vision 15 (NT) 59.3, 160.1, 203mr; Digital Vision 17 (NT) 120bl, 202tl; Don Farrall/Photodisc 29 (NT) 132bl, 238.1; Empics/Jerome Delay/AP Photos 33.3; Groupal Ltd 146.1; ICI 154.1; Illustrated London News V1 (NT) 113mr, 184.1; Imagin/London (NT) 26.2; Ingram ILS V2 CD 5 (NT) 134.3; Ingram PL V1 CD2 (NT) 7.2; James Lauritz/Digital Vision C (NT) 48.1; Karl Ammann/Digital Vision AA (NT) 122.1; Martyn F. Chillmaid 214.3, 242.3; Mary Evans Picture Library 264.1b; National Portrait Gallery 113b; Patrick Fullick 92tl, 137m, 137ml, 137tm; Perkin Elmer 267ml, 267tl; Peter Adams/Digital Vision BP (NT) 38.1; Photodisc 4 (NT) 100.2; Photodisc 17B (NT) 35m; Photodisc 18 (NT) 49.2; Photodisc 19 (NT) 82.1, 91.4, 255tl; Photodisc 22 (NT) 43.3, 62.2; Photodisc 29 (NT) 189.2; Photodisc 31 (NT) 18.3, 63.3; Photodisc 38A (NT) 202tr; Photodisc 44 (NT) 58.1, 168.2; Photodisc 60 (NT) 31.2; Photodisc 67 (NT) 76.1, 86.2a; Photodisc 71 (NT) 203tl, 255br; Photodisc 79 (NT) 35b; Pictoral Press/Alamy 157t; Pilkington Ltd 138tl; G T Woods 211.5; Rubberball WW (NT) 255bl; Science Photo Library 113tl, 145.2, 210.2a, 210.2b, 211.3, 248.1a, /Adrian Thomas 32.1, /Alan Sirulnikoff 47.3, /Andrew Lambert 20tr, 28.1, 72.2, 89.2, /Andrew Lambert Photography 196.1a, 196.1b, 196.2, 214.2, 226.1, 258.1, 258.2, 260.2, /Annabella bluesky 89.3, /Astrid & Hanns-Frieder Michler 239.3, /BSIP, Abbate 209ml, /CERN 127br, /Charles D Winters 24.1, /Chris Knapton 200.3, /Chris Sattlberger 138mr, /Claude Nuridsany and Marie Perennou 138ml, /CNRI 86.2b, /Cordelia Molloy 32.2, 72.1, 74.3, 79tl, 79tr, 82.2, 84.1, 84.2, 86.1, 88.1, 175.3, /David Parker 127tr, /David Scharf 101.4, /Dirk Wiersma 102.1, /Dr Jeremy Burgess 209tr, /Eye Of Science 138br, /GECO UK 265.2, /Geoff Kidd 202b, /Georgette Douwma 101.3, /J. C. Revy 266.1, /James Holmes/Celltech 267br, /James Holmes/Oxford Centre For Molecular Sciences 266bl, /James King-Holmes 126br, /John McLean 266ml, /Kaj R Svensson 70.1, /Lawrence Lawry 74.1, 136.2a, /Martin Bond 62.1, 208tl, /Martyn F. Chillmaid 4.1, 85.4, 262.1, /Maximilian Stock LTD 105.2, 162.2, 190.1, /Michael Donne 147.2, /Pascal Goetgheluck 45.2, /Pat & Tom Leeson 237.3, /Peter Menzel 138bl, /Robert Brook 63.4, /Russ Munn/AgstockUSA 208br, /Sheila Terry 126tl, 241.3, /Steve Allen 136.1, /Sue Baker 20ml, 166.1, /TEK Image 8.1, 264.1a, 267tr, /TRL LTD 54.1; StatOil 19.5; Steve Mason/Photodisc 46 (NT) 92br; Stockbyte 29 (NT) 75.4; Stockhaus UKBS (NT) 35br, 35ml; The Charcoal Burners Camp at the Weald & Downland Open Air Museum, Singleton, Chichester, West Sussex 34.1; Topfoto.co.uk, WDS 22tl; USDA 91.2

Picture research by Stuart Silvermore, Science Photo Library and johnbailey@ntlworld.com.

Every effort has been made to trace all the copyright holders, but if any have been overlooked the publisher will be pleased to make the necessary arrangements at the first opportunity.